Ingenious Genes

Life and Mind: Philosophical Issues in Biology and Psychology
Kim Sterelny and Robert A. Wilson, Series Editors

Ingenious Genes: How Gene Regulation Networks Evolve to Control Development, Roger Sansom, 2011

Yuck! The Nature and Moral Significance of Disgust, Daniel Kelly, 2011

Laws, Mind, and Free Will, Steven Horst, 2011

Humanity's End: Why We Should Reject Radical Enhancement, Nicholas Agar, 2010

Color Ontology and Color Science, Jonathan Cohen and Mohan Matthen, editors, 2010

The Extended Mind, Richard Menary, editor, 2010

The Native Mind and the Cultural Construction of Nature, Scott Atran and Douglas Medin, 2008

Describing Inner Experience? Proponent Meets Skeptic, Russell T. Hurlburt and Eric Schwitzgebel, 2007

Evolutionary Psychology as Maladapted Psychology, Robert C. Richardson, 2007

The Evolution of Morality, Richard Joyce, 2006

Evolution in Four Dimensions: Genetic, Epigenetic, Behavioral, and Symbolic Variation in the History of Life, Eva Jablonka and Marion J. Lamb, 2005

Molecular Models of Life: Philosophical Papers on Molecular Biology, Sahotra Sarkar, 2005

The Mind Incarnate, Lawrence A. Shapiro, 2004

Organisms and Artifacts: Design in Nature and Elsewhere, Tim Lewens, 2004

Seeing and Visualizing: It's Not What You Think, Zenon W. Pylyshyn, 2003

Evolution and Learning: The Baldwin Effect Reconsidered, Bruce H. Weber and David J. Depew, editors, 2003

The New Phrenology: The Limits of Localizing Cognitive Processes in the Brain, William R. Uttal, 2001

Cycles of Contingency: Developmental Systems and Evolution, Susan Oyama, Paul E. Griffiths, and Russell D. Gray, editors, 2001

Coherence in Thought and Action, Paul Thagard, 2000

Ingenious Genes

How Gene Regulation Networks Evolve to Control Development

Roger Sansom

A Bradford Book
The MIT Press
Cambridge, Massachusetts
London, England

For information about quantity discounts, email special_sales@mitpress.mit.edu.

Set in Stone Sans and Stone Serif by Toppan Best-set Premedia Limited.
Printed and bound in the United States of America.

Library of Congress Cataloging-in-Publication Data

Sansom, Roger.
Ingenious genes : how gene regulation networks evolve to control development / Roger Sansom.
p. ; cm.—(Life and mind)
"A Bradford book."
Includes bibliographical references and index.
ISBN 978-0-262-19581-2 (hardcover : alk. paper)
1. Genetic regulation—Computer simulation. I. Title II. Series: Life and mind.
[DNLM: 1. Gene Expression Regulation. 2. Gene Regulatory Networks.
3. Evolution, Molecular. QU 475]
QH450.S26 2011
572.8'65—dc22

2010053621

10 9 8 7 6 5 4 3 2 1

for Nana and Oupa

Contents

Preface ix
Acknowledgments xi

1 The Problem of the Evolvability of Gene Regulation Networks 1

2 Are Gene Regulation Networks Evolvable? 23

3 Kauffman's Framework for Gene Regulation 37

4 The Connectionist Framework for Gene Regulation 69

5 Why Gene Regulation Networks Are the Controllers of Development 97

Notes 113
References 115
Index 127

Preface

You are a collection of more than 10 trillion (10^{13}) cells, busy performing many tasks on which your continued existence depends. That is impressive enough, but consider too your history. You began as a single cell that repeatedly divided and differentiated to produce about a hundred cell types, as different as bone and neuron, arranged together in just the right way. This complex and delicate process may be commonplace, but it should be no less awe-inspiring for that. What is it that allows cells to pull off a trick as difficult as development?

Your genes are involved in the production of proteins that perform many cellular tasks. Most catalyze chemical reactions; some provide the building blocks for the rest of you; others work together to turn on and off the production of proteins made with other genes. These genes are called *transcription factors*. Their regulation of gene expression is crucial to overcoming the difficulty of development.

Your gene regulation network is contained in each cell, and the same network had to produce the right gene activity in the great variety of situations that the great variety of cells in your body have faced, all the way back to that first cell. Genes working together make up a truly ingenious system. How does the system work? How did it come about? This book proposes answers to those questions.

The first question is about how gene regulation occurs during development. How can a fundamentally simple system of genes turning each other on and off result in the right outcomes to such a variety of challenges? I propose that the key lies in how multiple transcription factors combine to regulate a simple gene. They act in a qualitatively consistent way. This means that each transcription factor is either always promoting the expression of a gene or always inhibiting that expression. This way of acting together is similar to the way neurons act in the brain to allow good outcomes to a variety of situations you face. Multiple transcription factors acting together in this way allow the expression of genes to be finely tuned to the variable microenvironments of cells.

How did you come to have such a refined simple system? Darwin's theory of evolution by natural selection is half of the answer to that question. Your ancestors' gene regulation networks accrued many adaptive mutations. Those mutations were unguided but lucky. The majority of unguided mutations to any system will do more harm than good, but some systems are virtually impossible to be improved in this way. I argue that gene regulation networks are quite the opposite. They are highly evolvable, with an impressively high chance of improvement by unguided mutation. The reason for this also boils down to the qualitative consistency of gene regulation networks, which allows mutations that add or remove transcription factors to have only small effects on the regulation of specific genes. Mutations that have small effects are more likely to be adaptive than mutations that have large effects. Therefore, the qualitative consistency of gene regulation networks allows them to be highly evolvable.

This book begins with some history of evolutionary theory, with a particular emphasis on explanations of organisms' adaptive complexity and evolvability. Much of the book concerns models or frameworks for thinking about gene regulation networks. Two frameworks for gene regulation networks are described in detail. The first was developed by Stuart Kauffman, who has argued that gene regulation networks are fundamentally systems that repeat patterns of gene expression. I criticize Kauffman's framework as inadequate to explain how gene regulation networks overcome the difficulty of development. I propose a second framework, which I call *the connectionist framework*. In so doing, I borrow work done in the fields of artificial intelligence and philosophy of mind and apply it to gene regulation. Finally, I try to bring this work to bear on how we should understand the developmental process that produced each of us. I propose that we gain insight when we identify gene regulation networks as the controller of development. This is due to both the nature and the evolution of development. The ingenuity of genes is explained by how gene regulation networks evolve to control development.

Acknowledgments

With thanks, I acknowledge the following people whose knowledge and insight contributed to this work while it was in development: Colin Allen, Michael Bergmann, Mark Bernstein, Martin Curd, Jeremy Draghi, Matthew Hahn, Adam Jones, Thomas McKnight, Dan McShea, John Roberts, John Scroggs, Kim Sterelny, Karola Stotz, Greg Wray, and the two referees for the MIT Press. I would like to especially thank Lindley Darden for her generous and thorough comments on the manuscript.

Some passages in chapters 3 and 4 appeared in "Countering Kauffman with connectionism: Two views of gene regulation and the fundamental nature of ontogeny" (*British Journal for the Philosophy of Science* 59, no. 2: 169–200), and some in "The connectionist framework for gene regulation" (*Biology and Philosophy* 23:475–491).

I'd like to thank my family, including Mindy Bergman, who helped by being an excellent writer and even better wife, and my son, Leo, who developed from a single cell into the billions and billions of cells that delight my days.

1 The Problem of the Evolvability of Gene Regulation Networks

Organisms are superbly well engineered to survive and reproduce in their environments. We humans have achieved a wide variety of engineering feats, from computers to space stations, and we have learned to manipulate natural organisms with biotechnology, but it remains well beyond our capacity to build from scratch even simple forms of life capable of surviving and reproducing in a natural environment. The engineering of complex organisms, such as vertebrates, probably will always remain beyond our reach. The only way we can produce a complex organism is the old-fashioned way.

Explaining the adaptive complexity of organisms has always been a goal of biology. It was long believed that the characteristics of species remained static and must have been designed by God. In fact, the superb engineering of organisms was the foundation of the argument from design for the existence of God. This argument was most famously presented by the natural theologian William Paley (1743–1805), who used the analogy of a watch and its maker:

> In crossing a heath, suppose I pitched my foot against a stone, and were asked how the stone came to be there; I might possibly answer, that, for anything I knew to the contrary, it had lain there forever: nor would it perhaps be very easy to show the absurdity of this answer. But suppose I had found a watch upon the ground, and it should be inquired how the watch happened to be in that place; I should hardly think of the answer I had before given, that for anything I knew, the watch might have always been there. . . . There must have existed, at some time, and at some place or other, an artificer or artificers, who formed [the watch] for the purpose which we find it actually to answer; who comprehended its construction, and designed its use. . . . Every indication of contrivance, every manifestation of design, which existed in the watch, exists in the works of nature; with the difference, on the side of nature, of being greater or more, and that in a degree which exceeds all computation. (Paley 1802)

If we can reason from the design of a watch to the existence of its designer, Paley argued, we can certainly reason from the design of an organism, which is finer than that of a watch "in a degree which exceeds all computation," to the existence of its designer—God.

A number of thinkers would challenge this conventional wisdom. I will introduce some of them here, not as elements of a complete history of evolutionary theory, but rather as elements of a narrative that leads to the problem this book addresses: the problem of the evolvability of gene regulation networks.

1.1 Erasmus Darwin

Erasmus Darwin (1731–1802)—physician, naturalist, and grandfather of Charles—proposed in his two-volume work *Zoonomia*, first published in 1794–1796, that every living organism on Earth had descended from a microscopic aquatic filament. All the various forms extant in Darwin's day had "transmuted" from this common ancestor. "In some this filament in its advance to maturity has acquired hands and fingers, as in mankind.," Darwin wrote (1801, p. 506). "In others it has acquired claws or talons, as in tygers and eagles. In others, toes with an intervening web, or membrane, as in seals and geese."

How can changes to a species come about, particularly changes that make an organism better suited to their environment by making them more complex? This is a major question for any evolutionary view of life. We would never expect pieces of metal to spontaneously come together to form a well-functioning complex system such as a watch. We would not even expect a damaged watch to spontaneously start working again. So how could organisms change from simple filaments to complex vertebrates?

Darwin saw the many aspects of the world improving slowly as an unfolding of God's plan. Society, geology, and even the cosmos progressed not because God regularly interfered, but because God had set up the laws of nature so as to result in constant improvement:

> Thus it would appear, that all nature exists in a state of perpetual improvement by laws impressed on the atoms of matter by the great CAUSE OF CAUSES; and that the world may still be in its infancy, and continue to improve FOR EVER AND EVER. (Darwin 1818, volume I, p. 437)

Darwin's view does provide an ultimate answer to the question of how adaptive complexity is gained, but it leaves open the question of how the laws of nature bring this change about. Darwin proposed two answers to that question. Most important, an organism is changed by its activity, and many of those changes are passed on to their offspring. And mate selection results in the strongest organisms' procreating and passing on the characteristics that made them strong:

> Fifthly, from the first rudiment, or primordium, to the termination of their lives, all animals undergo perpetual transformations; which are in part produced by their own exertions in consequence of their desires and aversions, of their pleasures and pains, or of irritations, or of associations; and many of these acquired forms, or propensities, are transmitted to their posterity.
>
> The birds, which do not carry food to their young, and do not therefore marry, are armed with spurs for the purpose of fighting for the exclusive possession of the females, as cocks and quails. It is certain that these weapons are not provided for their defence against other adversaries, because the females of these species are without armour. The final cause of this contest amongst the males seems to be, that the strongest and most active animal should propagate the species, which should thence become improved. (Darwin 1801, volume II, pp. 236–238)

Though some readers were enthusiastic about them, Darwin's ideas were largely dismissed as blasphemous (for proposing a view of nature in which God was not regularly and intimately involved) or highly speculative (for a lack of evidence). The poet Samuel Taylor Coleridge (1772–1834) coined the term "Darwinizing," which was "all surface and no content, all shell and no nut, all bark and no wood." Nora Barlow, editor of *The Autobiography of Charles Darwin: 1809–1882,* comments: "Fearing such reactions, Erasmus Darwin had withheld publication for twenty years. But this had not stopped him from adopting the family motto: *Ex omni conchis* (All from shells), which appeared at the bottom of the family coat of arms everywhere in the younger Darwin's childhood. His grandfather's advocacy of transmutation made him more open to develop his own theory, but he would ultimately also find *Zoonomia* too speculative."

1.2 Jean-Baptiste Lamarck

Jean-Baptiste Lamarck (1744–1829) was a French naturalist who, like Erasmus Darwin, wasn't shy about identifying and explaining "large facts" of nature. Lamarck proposed general materialistic theories of a wide range of chemical, meteorological, and geological phenomena. He viewed life as "nothing more than the movement in the parts of [organized] beings" (1797, p. 255). Though he didn't deny the "immortal soul" of man or the "perishable soul" of animals, Lamarck thought that, because they could not be known physically, they weren't proper concerns for science.

Lamarck shared Darwin's view that life was inherently progressive. In Lamarck's view, the "Sublime Author of nature" had created laws that meant that there was a plan of nature for life to become ever more complex:

> It is quite clear that both animal and vegetable organisation have, as a result of the *power of life*, worked out their own advancing complexity, beginning from that which was the simplest and going on to that which presents the highest complexity, the greatest number of organs, and the most numerous faculties; it is also quite clear that every special organ and the faculty based on it, once obtained, must continue to exist in all living bodies which come after those which possess it in the natural order.
>
> Everything is thus preserved in the established order. . . . Everywhere and always the will of the Sublime Author of nature and of everything that exists is invariably carried out. (Lamarck 1801, pp. 13–14)

The power of life necessarily pushed each lineage toward greater complexity, but there was also another element in the evolutionary process. The activity of organisms in response to their living conditions produced changes that could be passed on to the next generation. This accidental feature of life interrupted or modified the details of how lineages became more complex:

> It will in fact become clear that the state in which we find any animal, is, on the one hand, the result of the increasing complexity of organization tending to form a regular gradation; and on the other hand, of the inheritance of a multitude of very various conditions ever tending to destroy the regularity in the gradation of the increasing complexity of organization. (ibid., p. 16)

The first of these elements was more important to Lamarck than the second, but the second is more closely associated with him. Lamarck believed that if an organism frequently or consistently used an organ, that organ would strengthen or develop. Conversely, disuse of an organism would weaken the organ or cause it to deteriorate:

> . . . one may perceive that the bird of the shore, which does not at all like to swim, and which however needs to draw near the water to find its prey, will be continually exposed to sinking in the mud. Wishing to avoid immersing its body in the liquid, it acquires the habit of stretching and elongating its legs. The result of this for the generations of these birds that continue to live in this manner is that the individuals will find themselves elevated as on stilts, on long naked legs. (ibid. 1801, p. 4)

Lamarck even proposed a mechanism for this. Life was originally stimulated by the subtle fluids of heat and electricity. Their motion also formed "canals" that changed the organism:

> The characteristic of the movement of fluids in the supple parts of living bodies that contain them is to trace out routes and places for deposits and outlets; to create canals and the various organs, to vary these canals and organs according to the diversity of either the movements or the nature of the fluids causing them; finally, to enlarge, elongate, divide, and gradually solidify these canals and organs. . . . The state of organization in each living body has been formed little by little by the increasing influence of the movement of fluids and by the changes continually undergone in the nature and state of these fluids through the usual succession of losses and renewals. (1802, pp. 8–9)

And Lamarck believed that such changes were passed on to offspring:

> The law of nature by which new individuals receive all that has been acquired in organization during the lifetime of their parents is so true, so striking, so much attested by the facts, that there is no observer who has been unable to convince himself of its reality. (1815, volume 1, p. 200)

Today this "law" of the inheritance of acquired characteristics is primarily identified with "Lamarckianism." The identification is inappropriate for three reasons. First, that acquired characteristics were inherited was a common view in Lamarck's time. Second, Lamarck's most interesting contribution may have been his proposal as to how activity changed an organism (through the motions of fluids in canals) rather than his view that such changes were passed on to the next generation. Third, Lamarck thought that the power of life that necessarily pushed life toward greater complexity was more important to the evolutionary process than the inheritance of accidentally acquired characteristics.

Lamarck shared with Erasmus Darwin the view that nature's tendency to become more complex, which had been granted by the "Sublime Author" or the "Cause of Causes," was fundamental to evolution. The two men also agreed that acquired characteristics were inherited. There was a crucial difference between their theories, however. Darwin believed that all animals were related, and saw a creative order in the battle for procreation. Lamarck, in contrast, thought that each lineage was so imbued with the power of life that it would inevitably become more complex with time. Why do we see both complex life and simple life? Lamarck believed that the presence of complex life shows that this process has had time to improve life. He attributed the presence of simple life to nature's recurrently "directly" generating the simplest forms of life "with facility" under "favorable circumstances" (1801, p. 41). Thus, the world was even more inherently progressive to Lamarck than it was to Darwin. On Lamarck's view, organisms were so driven to greater complexity that simple life must be relatively new.

1.3 Charles Darwin

Charles Darwin (1809–1882) read his late grandfather's work as a young man, and was exposed to Lamarck's ideas at the University in Edinburgh. Later he characterized Lamarck as "the first man whose conclusions on the subject excited much attention" (Darwin 1861, p. xiii). "This justly celebrated naturalist," Darwin continued, "did the eminent service of arousing attention to the probability of all changes in the organic, as well as in the inorganic world, being the result of law, and not of miraculous interposition."

In 1831, Darwin departed England on a survey ship, *H.M.S. Beagle*, as a naturalist and as a dining companion for Captain Robert FitzRoy. In the course of the *Beagle*'s round-the-world voyage, which took nearly five years, Darwin had an extensive opportunity to investigate the biology of South America and the Galapagos islands. "During the voyage of the *Beagle*," he wrote in his autobiography,

> I had been deeply impressed by discovering in the Pampean formation great fossil animals covered with armour like that on the existing armadillos; secondly, by the manner in which closely allied animals replace one another in proceeding southwards over the Continent; and thirdly, by the South American character of most of the productions of the Galapagos archipelago, and more especially by the manner in which they differ slightly on each island of the group; none of these islands appearing to be very ancient in a geological sense. (Barlow 1958, p. 118)

Darwin returned puzzled by the distribution of biological life that he had observed. Fossils he had unearthed in South America looked more similar to that continent's present inhabitants than any other known life forms did. It had long been observed that organisms were well suited to their environments, but Darwin did not know why they were similar to other local species inhabiting different environments. Darwin saw that many species found only on a single island in the Galapagos were similar to species found on the nearby South American continent. It was as if they had traveled from South America to the Galapagos and then been modified to suit the local environment. "It was evident," Darwin wrote in his autobiography,

> that such facts as these, as well as many others, could be explained on the supposition that species gradually become modified; and the subject haunted me. But it was equally evident that neither the action of the surrounding conditions, nor the will of the organisms (especially in the case of plants), could account for the innumerable cases in which organisms of every kind are beautifully adapted to their habits of life,—for instance, a woodpecker or tree-frog to climb trees, or a seed for dispersal by hooks or plumes. I had always been much struck by such adaptations, and until these could be explained it seemed to me almost useless to endeavour to prove by indirect evidence that species have been modified. (Barlow 1958, pp. 118–119)

Among the books Darwin took with him on the *Beagle* expedition was Charles Lyell's *Principles of Geology* (1830). Darwin was impressed with Lyell's gradualist explanations, which relied on cumulative change over a long period of time rather than on sudden catastrophic events.

The final piece of Darwin's puzzle fell into place in 1838 when he read *An Essay on the Principle of Population*, published in 1798 by Thomas Robert Malthus (1766–1834), a minister of the Church of England who later became a professor of modern history and political economy. Malthus theorized that, if left unchecked, the human population would increase geometrically and would eventually be unsustainable. It occurred to Darwin that animal populations too, left unchecked, would increase geometrically. From the fact that they hadn't become unsustainable, he concluded that a great many organisms were destroyed before they could reproduce. Darwin then became interested in the capacity of selective breeding to produce new varieties, because he wondered if the differential survival of variants might ultimately lead to the creation of new species. Later he would write:

> As many more individuals of each species are born than can possibly survive; and as, consequently, there is a frequently recurring struggle for existence, it follows that

any being, if it vary however slightly in any manner profitable to itself, under the complex and sometimes varying conditions of life, will have a better chance of surviving, and thus be naturally selected. From the strong principle of inheritance, any selected variety will tend to propagate its new and modified form. (Darwin 1859, p. 5)

To avoid a negative reaction such as those that had greeted Lamarck's and his grandfather's respective theories of transmutation of species, Darwin dedicated himself to amassing evidence for his view before going public with it. He spent eight years producing a classification of barnacles. Although he had evolutionary theory in mind, and was trying to imagine the characteristics of the common ancestor of all barnacles, his published results (1851, 1854) did not mention evolution explicitly. Darwin acknowledged that he published *On the Origin of Species by Means of Natural Selection, or The Preservation of Favored Races in the Struggle for Life* as soon as he did (1859) only because Alfred Russel Wallace (1823–1913)—another naturalist who had been inspired by Malthus—was about to propose a similar theory, which he described in an essay mailed to Darwin in 1858.

Charles Darwin's fundamental difference with his grandfather and with Lamarck was that he did not see nature as an inevitably progressive enterprise. He wrote to his botanist friend D. J. Hooker: "Heaven forefend me from Lamarck['s] nonsense of a 'tendency to progression', 'adaptations from the slow willings of animals,' etc.!" (1844). Rather than the unfolding of God's plan, evolution by natural selection was a messy and wasteful business. Indeed, not only was evolution not inherently progressive; it was not law-like in any widely recognized way. The direction of evolutionary change was left to the vagaries of changes in the environment. Before the publication of *The Origin of Species,* Darwin put it this way in another letter to Hooker: "What a book a Devil's Chaplain might write on the clumsy, wasteful, blundering, low and horribly cruel works of nature." (1856)

On Darwin's view, explaining the evolution of complexity was both easier and harder than it was for his grandfather and Lamarck. It was easier because changes that increased adaptive complexity could be quite rare on his view. Unlike Lamarck, Darwin did not propose that a population would become better adapted because each individual organism was likely to produce better-adapted offspring. Instead, a population may increase in adaptive complexity because adaptive variations, even if rare, can be spread through the population by natural selection. It was harder because Darwin did not propose that God had set things up for the rise of adaptive complexity, or that the efforts of organisms generally lead to adaptive traits in their offspring.

Darwin admitted that it was hard to imagine how evolution by natural selection could result in the "organs of extreme perfection" that we see in nature, but did not think that was sufficient reason to reject the theory:

> To suppose that the eye, with all its inimitable contrivances for adjusting the focus to different distances, for admitting different amounts of light, and for the correction of spherical and chromatic aberration, could have been formed by natural selection, seems, I freely confess, absurd in the highest possible degree. Yet reason tells me, that if numerous gradations from a perfect and complex eye to one very imperfect and simple, each grade being useful to its possessor, can be shown to exist; if further, the eye does vary ever so slightly, and the variations be inherited, which is certainly the case; and if any variation or modification in the organ be ever useful to an animal under changing conditions of life, then the difficulty of believing that a perfect and complex eye could be formed by natural selection, though insuperable by our imagination, can hardly be considered real. (1859, pp. 186–187)

Darwin thought that highly complex well-functioning organisms could have resulted from gradual variation. He intuitively grasped the idea that a random change is never likely to improve a well-designed system, but that the probability of improvement increases if the change is small. Natural selection could accumulate many rare small changes to evolve superb engineering with enough time:

> Almost every part of every organic being is so beautifully related to its complex conditions for life that it seems as improbable that any part should have been suddenly produced perfect, as that a complex machine should have been invented by man in a perfect state. (1861, p. 46)

The vast array of examples that Darwin compiled, and the many objections to his views and appropriate responses that he considered, showed a remarkable understanding of his theory. Still, there were gaps in the knowledge of his day, and one of these gaps led to an objection that it could not account for the evolution of complexity.

The most important acknowledged unknown in evolutionary theory in Darwin's time was the mechanism of inheritance. Darwin accepted the dominant contemporary views of inheritance: pangenesis (i.e., that corpuscles of inherited material were throughout the body) and blended inheritance (that these corpuscles from both parents were blended together to produce offspring). On these views, an offspring is typically the average of its parents. A child of one short parent and one tall parent would be likely to be of average height, and its offspring would also likely be a blend of its parents' heights. But the "blending view" of inheritance leads to a problem for the theory of evolution by natural selection, as was pointed out in an 1867 letter to Darwin by Fleeming Jenkin, an engineer at the University of Edinburgh: Any adaptive character that appears in a member of a population will be incrementally blended away over the generations as individuals with that character breed with individuals without it. This left Darwin looking for something to supplement his theory and to provide more variations that increased adaptive complexity. For a lack of an alternative, Darwin turned to the inheritance of acquired characteristics. He had considered it unimportant in the first edition of *Origin*, but he gave it greater importance in subsequent editions.

Though the theory of evolution by natural selection had widespread immediate influence, it faced constant criticism. As late as 1900, natural selection was not considered the most important factor in evolution.[1]

1.4 Gregor Mendel

While Darwin was developing his theory of evolution, Gregor Mendel (1822–1884), an Augustinian monk in what is now the Czech Republic, was performing experiments that would ultimately lead to the rejection of Darwin's blending view of inheritance and Lamarck's view of the inheritance of acquired characteristics.

Mendel was interested in variation within species. He found varieties of pea plants that consistently yielded similar offspring characteristics, such as seed color, which varied among the rest of the population. He then crossed these varieties with each other. He found that with regard to certain characteristics, the next generation did not have an intermediate form, but instead had the form of one of its parents. For example, the offspring of one yellow-seeded and one green-seeded parent were all yellow. He called yellow "dominant" and green "recessive." The same character was dominant, whether it came from the male or the female parent. When these hybrids were then self-pollinated, the ratio of dominant to recessive characters was 3:1, on average. Those exhibiting the recessive character bred true (i.e., when self-pollinated produced only the recessive character). One-third of those exhibiting the dominant trait also bred true. Thus, there was a ratio of 2:1:1 (mixed: pure dominant: pure recessive). Mendel put it this way: "*It is now clear that the hybrids form seeds having one or other of the two differentiating characters, and of these one-half develop again the hybrid form, while the other half yield plants which remain constant and receive the dominant or the recessive characters in equal numbers.*" (source: C. T. Druery and William Bateson, "Experiments in plant hybridization," translation of Mendel's 1865 paper, *Journal of the Royal Horticultural Society* 26, 1901: 1–32) This insight led to what became known as the First Law of Mendelism.

Mendel also crossed hybrids that did not breed true in one trait with hybrids that did not breed true in another. He found that the ratios for each character continued to apply, independent of the other characters. "There is therefore no doubt," he wrote, "that for the whole of the characters involved in the experiments the principle applies that *the offspring of the hybrids in which several essentially different characters are combined exhibit the terms of a series of combinations, in which the developmental series for each pair of differentiating characters are united.* It is demonstrated at the same time that *the relation of each pair of different characters in hybrid union is independent of the other differences in the two original parental stocks.* (ibid.) This insight led to what became known as the Second Law of Mendelism.

1.5 William Bateson

Mendel's original paper, "Versuche über Pflanzenhybriden," was first published in *Verhandlungen des naturforschenden Vereines in Brünn* (Proceedings of the Brünn Natural History Society) in 1866, but was largely ignored until 1900, when it was rediscovered by Hugo de Vries and Carl Correns. William Bateson (1861–1926), who also read it in 1900, had it translated in 1901 and became a staunch advocate for Mendelism. Bateson would go on to name the study of inheritance "genetics" and to found (in 1910) the *Journal of Genetics*.

Bateson already had been studying heredity and had published (in 1894) a book titled *Materials for the Study of Variation: Treated with Special Regard to Discontinuity in the Origin of Species*. Bateson studied a number of lakes in Kazakhstan that differed in salinity. Together, they provided an example of continuous variation in salt levels across environments. Darwinists of Bateson's day would expect to find continuous variation in species across such a continuous variation in environments, but Bateson found very little variation at all, and what variation he did find was discontinuous. Accordingly, while he accepted natural selection occurred, he held that it worked on limited and discontinuous variation, and that the source of this variation was more important in evolutionary explanation than natural selection was. This was an example of mutation theory. Bateson's work had come under intense criticism from Darwinists, but Mendelism supported mutation theory over the conventional Darwinism of the day, which was natural selection of blended inheritance.

Mutation theory made mutations central to evolutionary change. Bateson considered John P. Lotsy's (1867–1931) position that all variation was due to recombination of Mendelian factors, but he thought there were counter-examples to this view. One example was the "Crimson King" variety of *Primula*, which bred true for many generations before producing the "Salmon King" variety, which differed in the loss of one color factor and which also bred true. Bateson considered this novelty to be the novelty of loss. Bateson attributed other instances of variation (e.g., the picotee sweet pea, whose flower has only purple edges, rather than being completely purple) to the fractionation of factors. The factor responsible for purple flowers was fractured by irregularities in segregation. Bateson thought this similar to irregularities in development, such as when leaves are partially petaloid. These are cases where something inherited is damaged or lost. However, Bateson could not fathom the addition of inherited factors or mutations that improved their function. This attempt to understand the evolution of adaptive complexity led him to suggest an extraordinary view of evolution in his presidential address at the Australian meeting of the British Association for the Advancement of Science in 1914:

> Having in view these and other considerations which might be developed, I feel no reasonable doubt that though we may have to forego a claim to variations by addition of factors, yet variation both by loss of factors and by fractionation of factors is a genuine phenomenon of contemporary nature. If then we have to dispense, as seems likely, with any addition from without we must begin seriously to consider whether the course of evolution can at all reasonably be represented as an unpacking of an original complex which contained within itself the whole range of diversity which living things present. I do not suggest that we should come to a judgment as to what is or is not probable in these respects. As I have said already, this is no time for devising theories of evolution, and I propound none. But as we

have got to recognize that there has been an evolution, that somehow or other the forms of life have arisen from fewer forms, we may as well see whether we are limited to the old view that evolutionary progress is from the simple to the complex, and whether after all it is conceivable that the process was the other way about." (1914, p. 298)

To clarify the idea Bateson describes here, it is necessary to distinguish developmental complexity from genetic complexity. A quick glance at a bacterium and a mammal shows that the mammal has greater adaptive complexity. This complexity is measured in terms of how many types of parts make up an organism and how different those parts are from each other. This complexity is reproduced anew in individual organisms. It concerns the development of the individual organism. Even if bacteria do not grow larger and change during their lifetime, they still undergo development in the limited sense of keeping themselves alive. However, they have less adaptive complexity (or, as it may be more precisely called, adaptive developmental complexity).

We can also ask questions about the genetic complexity of the organism. This is a measure of the factors that have survived countless generations and are passed on from parents to offspring. We might expect the genetic inheritance of bacteria to be simpler than the genetic inheritance of vertebrates, but this is an empirical issue. In order to know which has the greater genetic complexity, one must know how genetic factors work so that one can then appropriately measure the genetic complexity of individuals. Bateson did not have access to this information. However, he had observed loss of characteristics that suggested loss of genetic factors. He had examples of the environment's producing change, such as the presence of iron's turning a hydrangea pink, but this influence is not passed on to the next generation. Mostly, he could not conceive how environmental factors could become integrated within an organism in a way necessary for it to be grown by the organism to allow those factors to be passed on to individuals of future generations. Instead, he could only think of how a descendent might be missing the genetic factors responsible for the inheritance of a capacity to deal with its environment in a certain way.

We should acknowledge, as Bateson did, the lack of knowledge about the mechanisms of inheritance in his time. Mendel provided a very general description of the mechanism based on his observation of patterns of character inheritance, but no details about genes. This is particularly relevant because acquisition or loss of a factor may not be readily apparent when observing characteristics, because the acquisition of a factor might have the same effect on characteristics as the loss of an inhibiting factor.

On the view that Bateson described in the passage quoted above, the only way mutation of genetic factors could result in added developmental complexity is by the loss of inhibiting factors. Bateson validly concluded that *if* genetic complexity could not be added by mutation, then all genetic complexity must have been somehow stored in life's original forms. Even mutations that increase developmental complexity reduce genetic complexity. Thus, mutation was always, in a sense, an "unpacking" of what was there at the origin of life, genetically speaking. Bateson accepted that this was a non-intuitive position that turned evolution on its head. Evolution no longer took life from the genetically simple to the complex, as it inevitably did according to Lamarck and as it could according to Darwin; now it took life from the genetically complex to the simple.

Bateson's idea is a radical answer to the question of the evolution of adaptive complexity. Erasmus Darwin and Lamarck proposed that God had imbued nature with a propensity to improve, and that adaptive complexity therefore was an indirect result of the work of God. Charles Darwin proposed that God had done no such thing, and that nature was not universally driven to improvement, although improvement sometimes occurred and the influence of those occasions might be magnified by natural selection. Adaptive complexity was the occasional consequence of a chancy, messy, wasteful, and entirely natural system. This radical idea captured the public's imagination.

Bateson's proposal was in one respect a throwback to the orthogenesis of Erasmus Darwin and Lamarck, because all life became a lawful unfolding of what was there at the origin. In general, though, Bateson's proposal was extremely radical. According to it, life, when it began, was developmentally simple, but genetically so complex that all the genetic factors necessary for all life was already there. Not only was this idea non-intuitive; it was also implausible, because it left unanswered the question of how genetic complexity—necessary for the evolutionary future—could have been present at life's origin.

Just how many components the first reproducing organism had and how precisely they had to be put together for reproduction to occur determine the probability that the system's components came together spontaneously. All things being equal, the more precise the arrangement of each component must be, and the more components required, the less likely it is that the components will come together spontaneously. Bateson's proposal that the first reproducer contained all the genetic complexity of subsequent forms of life suggests that it had more genetic components than those who accept that the first reproducer was relatively simple. Thus, the origin of life looks vastly less probable on Bateson's scenario than on a Darwinian scenario.

Another reason why Bateson's speculation is so interesting is that it is an early attempt to figure out how the evolution of genetic inheritance is related to the evolution of organisms as a whole. In 1883, August Weismann (1834–1914) argued that in nature there is a fundamental distinction between the sperm or egg cells (the germ line), which carry factors that may be inherited by future generations, and other cells of the body (the soma), which do not. In one famous experiment, Weismann cut off the tails from mice for several generations, but the mice continued to have offspring that grew tails. This insight, which put pressure on the idea that acquired characteristics were inherited, would go on to shape the evolutionary theory of the twentieth century—perhaps more than any insight other than Darwin's theory of evolution by natural selection.

1.6 The Rise of the Gene

Bateson's speculation raised the question of how mutation could increase genetic complexity, but it was too radical to receive much attention. However, Bateson's use of Mendelism to criticize Darwinism required a response. It came from Ronald Fisher (1890–1962) when he wrote *Genetical Theory of Natural Selection* (1930), which became the foundation of evolutionary population genetics. He showed, quite contrary to Bateson, that not only was Mendelian genetic inheritance compatible with effective natural selection, but selection was more effective with Mendelian inheritance than with the blending inheritance that Charles Darwin accepted. In fact, in view of the rates of mutation observed, inheritance of discrete genes, rather than the blended inheritance accepted by Darwin, was required for effective evolution by natural selection. Mendelism solves Fleeming Jenkin's problem because Mendelian characters are not necessarily blended away over the generations as individuals with them breed with other individuals. Julian Huxley (1942) called the integration of Darwinism, Mendelism, and population genetics "the modern synthesis."

Advances in Mendelian genetics and the modern synthesis were followed by rapid advances in genetics. George Beadle and Edward Tatum (1941) argued that each gene corresponds to one "primary" character and one enzyme. Oswald Avery, Colin MacLeod, and Maclyn McCarty (1944) suggested that DNA was the basis of Mendelian heredity. James Watson and Francis Crick (1953) utilized the work of Rosalind Franklin to identify the structure of the DNA molecule, and Crick (1958) laid out the "central dogma" of the relationship between nucleic acids and protein, which contributed to the breaking of the genetic code in the 1960s.

With this gain in knowledge of the mechanisms of genetic inheritance came a straightforward explanation of how complexity could be added to the genome. We now know that genetic material can be added by mutation. Bateson simply, and perhaps understandably, failed to imagine the possibility of duplication of genes, for example. In general it is probably unwise to argue from being unable to imagine how life could do something to arguing that it does not. To his credit, Bateson did not claim to have proof for his radical view; rather, he wanted to pave the way for its conceivability, to allow the possibility of its later acceptance.

Thus, evolutionary theory includes a simple answer to the question of genetic complexity, but the question of adaptive developmental complexity remains. This is a question of how genetic changes can lead to adaptive changes that increase the complexity of development. In the middle of the twentieth century, perhaps for the first time in the history of evolutionary theory, the question of adaptive developmental complexity ceased to be the focus of attention for evolutionists. With advances in genetics, and especially in population genetics, evolution became widely treated by researchers as a change in the genomes of a population over time. This way of thinking culminated in Richard Dawkins's 1976 book *The Selfish Gene*. According to Dawkins, the gene is the fundamental unit of natural selection (ibid., p. 11) because genes can replicate accurately over many generations. Furthermore,

> Natural selection in its most general form means the differential survival of entities. Some entities live and others die but, in order for this selective death to have any impact on the world, an additional condition must be met. Each entity must exist

in the form of lots of copies, and at least some of the entities must be *potentially* capable of surviving—in the form of copies—for a significant period of evolutionary time. Small genetic units have these properties: individuals, groups, and species do not. (ibid., p. 33)

And, according to Dawkins,

A gene can live for a million years, but many new genes do not even make it past the first generation. The few new ones that succeed do so partly because they are lucky, but mainly because they have what it takes, and that means they are good at making survival machines. They have an effect on the embryonic development of each successive body in which they find themselves, such that that body is a little bit more likely to live and reproduce than it would have been under the influence of the rival gene or allele. For example, a 'good' gene might ensure its survival by tending to endow the successive bodies it finds itself in with long legs, which help those bodies escape from predators. (ibid., p. 36)

Natural selection selects genes typically, but not always, by selecting adaptive organisms. This view simply assumes that mutated genes can add adaptive complexity. The process of development, complex or not, may be considered a black box in the process of evolution, because the development of an organism with various traits is just a mechanism to differentiate the fitness of genes. If one has any doubts as to how extremely gene-focused Dawkins takes his views to be, one need only look at the first paragraph of the preface to *The Selfish Gene*:

This book should be read almost as though it were science fiction. It is designed to appeal to the imagination. But it is not science fiction: it is science. Cliché or not, "stranger than fiction" expresses exactly how I feel about the truth. We are survival machines—robot vehicles blindly programmed to preserve the selfish molecules known as genes. This is a truth which still fills me with astonishment. Though I have known it for years, I never seem to get fully used to it. One of my hopes is that I may have some success in astonishing others. (p. v)

In the twentieth century, the conceptual separation of evolution from development was widely seen as an important step in the rapid advance of biology because it allowed evolution to be studied without getting bogged down in the complexities of development. Developmental biology continued to be studied, but, that study was largely separated from the study of evolution (although it did assume various general ideas about adaptation and common descent). This segregation has always had dissenters, however (Burian 1986; Laubichler and Maienschein 2007). For example, in criticizing Dawkins's view, Stephen Jay Gould (2001, p. 213) argued that "units of selection must be actors within the guts of the mechanism, not items in the calculus of results.".

There is a tradeoff between the oversimplification of focusing on the evolution of genes and the daunting prospect of understanding the evolution of the entire developmental process that is the life of an organism. In this book, I shall attempt to go one step beyond the evolution of genes toward the evolution of organisms by offering a general theory of the first level of development beyond the gene: gene regulation networks.

1.7 Evolvability

The view that evolution is the change in distribution of genes in a population does not explain adaptive complexity. Indeed, such a view is blind to developmental complexity, because development doesn't enter into the picture at all. Dawkins believes that adaptive complexity is achieved, as Darwin proposed, by mutations' making small changes that are accumulated by natural selection. Dawkins (1996) even recounts the history of the small physiological changes that took place in the evolution of the eye, for example. However, this story leaves out an explanation of how gradual mutation is possible.

Richard Lewontin (1978) set out to give a general explanation of the evolution of adaptive complexity by pointing out the requirements it puts on the mutation of organisms. He proposed that only "adaptive evolution" can lead to "organisms as we know them," and that it can work only on organisms whose traits are "quasi-independent" and "continuous." Quasi-independent traits change independent of other traits. This allows natural selection to change populations trait by trait, rather than mutations that improve one trait also affecting other traits in ways that are likely to be maladaptive. Traits must also change continuously: small changes in a trait must have only small effects on fitness. Lewontin claimed that "continuity and quasi-independence are the most fundamental characteristics to the evolutionary process" (p. 169). Lewontin is right that the evolution that led to the complex adaptive organisms we know required quasi-independence and continuity, which are specific ways of achieving gradual mutation. However, neither Richard Lewontin nor Richard Dawkins nor Charles Darwin explained what it is about organisms that allows them to evolve gradually.

Dawkins (1989) coined the term "evolvability" in reporting an investigation that involved simulated life. Evolvability is a propensity to mutate adaptively. (For an analysis, see Sansom 2008c.) The philosopher Kim Sterelny takes the explanation of adaptive diversity to be the most important task for evolutionary theory and accepts the theory of evolution by natural selection to be a crucial part of the answer. (See Sterelny and Griffiths 1999.) Faced with the fact that natural selection will be effective only on systems that are sufficiently evolvable, Sterelny, for one, is willing to accept that the most important remaining question in evolutionary biology is how organisms can mutate adaptively at the rate (however low) necessary for natural selection to result in diverse forms of life that are so well adapted for reproduction in their environments (Sterelny 2000).

Mutations that generate adaptive complexity are the hardest type of mutation to explain. Mutations are thought to be random in the sense that they are not made any more likely simply by the fact that they would be adaptive. If one makes such a random change to a well-designed system, that change probably will produce no change in the effectiveness of that system to perform a task or will remove some necessary capacity of the system. But how can a random change result in the addition of a capacity? The performance of complex systems, such as electronic or mechanical human artifacts, is not likely to be improved by a random change, such as that induced by a swift kick (the practice of many frustrated users). Organisms are more advanced pieces of engineering than our electronic devices, so how can a random change to an organism (a metaphorical swift kick to the genome) add adaptive complexity?

It is pretty easy to imagine how an organism might lose a capacity with mutation. A change could prevent a system from functioning effectively, just as kicking a functioning television might break it. It is even easy to see how such a loss might be adaptive, if performing that function costs energy and it does not contribute to the organism's adaptivity in its environment (although it might have been adaptive in the environments of its ancestors). But how could a mutation allow a new component to be built at all, let alone be functional and adaptive? Kicking a television will not allow it to pick up radio signals.

One possible response is to provide almost no solution to the problem at all. One might accept that mutations that add adaptive complexity are vanishingly rare. A mutation's adding adaptive complexity to an organism may be no more probable than a random change's adding a new capacity to your television, for example. This response continues as follows: A vanishingly rare possibility is sufficient for natural selection to result in the complex adaptivity we see today, because populations of sufficient size and fecundity have existed long enough on Earth. Such a response may be consistent with much of what is claimed by Dawkins, who grants virtually all explanatory power to natural selection. If the evolution of adaptively complex organisms is the most important explanandum in evolutionary theory, and if organisms are vanishingly unlikely to produce mutants with increased adaptive complexity, and natural selection results in the evolution of adaptively complex organisms anyway, then natural selection is worthy of the amount of explanatory power that Dawkins grants it. However, if there are features of the way organisms mutate that add to the likelihood of mutations' increasing adaptive complexity, then those features deserve to be elements in an explanation of the adaptive complexity that we find in life.

Though the non-answer may lead a confirmed evolutionist (such as Dawkins) to overestimate the explanatory importance of natural selection, it may also provide support for skepticism about evolution. The best-known current alternative to evolutionary theory is Intelligent Design. All its arguments are arguments for the unevolvability of complex organisms. Michael Behe claims that certain systems of organisms are "irreducibly complex." Such systems have many well-designed components that are essential to the function of the system as a whole (1996, p. 39). The argument runs that natural selection could not design systems that are irreducibly complex, because the system would only be minimally functional once most of the designing had already been achieved. Before this point was reached, improvements in design would have no functional advantage, because they would still be completely dysfunctional and so would offer no selective advantage. Therefore, Behe reasons, that such systems "show strong evidence of design—purposeful, intentional design by an intelligent agent." (ibid., p. 98)

There are many responses to Intelligent Design arguments. Some have argued that the systems to which Behe points are not irreducibly complex, and have noted that there are simpler forms of those systems that still carry out the same functions. Others have argued that the ancestor system carried out a different function, and that thus the claim of irreducible complexity (which must be made relative to a particular function) is not relevant to showing that the system could not have evolved by natural selection (for another function). What is at issue in this debate is how we are to understand the evolvability of organisms. Charles Darwin anticipated such objections in his discussion of organs of extreme perfection (cited in section 1.3), and they have been made in various forms ever since Darwin's day. They are yet another illustration of the centrality of evolvability to evolutionary theory.

Lewontin proposed that quasi-independence and continuity are requirements for adaptive evolution. There may be other characteristics that are necessary for mutation to lead to novel features in development. To explain how organisms satisfy these requirements is to explain evolvability, which is to answer what Sterelny takes to be the big remaining question of evolutionary biology. Growing interest in the relationship between evolution and development has resulted in a growing field of theoretical and empirical research called evolutionary developmental biology ("evo-devo"). Any contribution to solving the problem of evolvability is firmly within this field. I hope that this book is such a contribution.

Some have suggested that evolutionary developmental biology requires a drastic reconceptualization of evolutionary theory. (See, e.g., Arthur 2000; Robert 2004.) After all, there is more to development than the activity of genes. For example, there are homeostatic and developmental processes that operate at levels higher than gene expression, such as at the tissue level, organ level, and so on. A more conservative approach is to try to extend a gene-centered view into development. This requires that we understand development as best we can, working our way up from the level of genes. This book is a contribution to this more conservative strategy. It offers an analysis of what I take to be one level up from individual genes: gene regulation networks (described in the next chapter). Such an approach will face the criticism that it is not the radical reconceptualization that is required. It may even be seen by some as a backward step, because it could stifle the radical agenda by perpetuating the myth that we can understand evolution from the gene level up. I will argue why gene regulation networks are of particular importance in understanding evolution in the final chapter, but, at this point I will simply state that I choose the strategy of building on what we have.

Concentrating on gene regulation networks is less reductive than gene selectionism, but it is still reductionistic. Accordingly, it has the potential to offer the strengths and the weaknesses of reductive analyses, which have been fruitful in science yet which never tell the whole story. Their strengths and weaknesses are both due to the fact that they simplify the problem. If successful, this project might set the stage for the next stage in the conservative gene-centered approach, which will tell more of the story. This project requires explaining quasi-independence and continuity in gene regulation networks. There will remain questions of exactly how these will relate to full-blown phenotypic traits, but the longest journey begins with one step.

2 Are Gene Regulation Networks Evolvable?

"Exactly how does an egg produce legs, head, eyes, intestine, and get up and start running about?" With this question, Conrad Waddington (1966, p. iv) expresses wonder at both any system that is capable of overcoming the difficulty of development and the difficulty of understanding such a system. The difficulty of a task can be understood in terms of the range of possible physical systems that can overcome that task. For example, opening a particular lock is difficult, because only a relatively small range of systems have the necessary features to be a key that can be inserted and turned to open that lock. Opening many different locks is more difficult. A smaller range of systems are skeleton keys. In contrast, the range of systems that will make an adequate paperweight is much broader. The requirements of moderate size and solidity are easily satisfied. Many of the various things lying on my desk would make an adequate paperweight, but none of them will open a lock.

Some of the difficulty of development lies in the fact that, to develop, an organism must do different things at different times or in response to changes in its environment. For example, *E. coli* will consume lactose only if there is no glucose available (Reznikoff 1992). In addition, cells in organisms face different microenvironments within the organism and must change their activities accordingly.[1] This happens most dramatically during the early development of complex organisms. Each of us humans began as a single cell. That zygote underwent about fifty cell divisions to form an infant with 10 trillion cells. Those cells differentiated from one another to form about 100 cell types. The cells differ not because they have different genes inside them (most cells share very similar DNA, and most cell differences are not explained by differences in primary DNA sequence), but differences in the expression of genes instead (i.e., which genes are transcribed to produce proteins). The gene regulation system must be one of the most ingenious systems in the body, because it must solve one of the most challenging difficulties of development.

2.1 Gene Regulation Networks

The term "gene" has multiple definitions in evolutionary biology (Griffiths and Neumann-Held 1999). In this book, I take a gene to be a section of DNA that is directly utilized in the production of a messenger RNA transcript, which is then used to synthesize a particular protein. A gene has two sections. The transcribed region is used as a template to form messenger RNA in the transcription process. The messenger RNA is then used to produce a protein in the translation process. Whether or not the transcribed region of each gene is transcribed is determined by the interaction of transcription factors and the regulatory region of the gene. The regulatory region has binding sites that specific transcription factors may bind to and thereby either activate or repress the transcription of the gene. In eukaryotes, the regulatory region of a gene need not be adjacent to its transcribed region and may be involved in regulating multiple genes. Transcription factors are themselves proteins that are produced by transcription and translation of a subset of genes. Thus, the expression of some genes can influence the expression of others. The relationship between the expression of genes of an organism (i.e., the expression of which gene activates or represses the expression of which other gene) makes up the organism's gene regulation network. Gene expression is crucial to many cellular activities. We can ascribe a significant aspect of the difficulty of development of the organism to the difficulty faced by the gene regulation network in expressing adaptive gene products for each microenvironment faced by each cell over the life of the organism.

Though understanding this gene regulation network is not all that is required to understanding biological development, it is a crucial step (Wright 1982; Raff 1996; Wilkins 2002). Even Waddington, who was well aware of levels of organization above gene regulation, such as structures inside cells, tissues consisting of many cells, organs, and finally the body as a whole, still endorsed Thomas Hunt Morgan's (1866–1945) view "that the fundamental agents that bring about embryonic development are the genes, and the only finally satisfactory theory of embryology must be a theory of how the activities of genes are controlled" (Waddington 1966, p. 18). This view requires a theory that explains the adaptivity of gene regulation networks.

In view of natural selection's capacity for favoring adaptive mutations, knowing how gene regulation networks are evolvable is important to an understanding of the ingenuity of gene regulation. How can gene regulation networks mutate in a continuous and quasi-independent manner? The direct outcome of gene regulation is the expression profile of each gene. This is a description of the microenvironments in which that gene is transcribed and the microenvironments in which it is not transcribed. In general, we expect that, if mutation to gene regulation is to be continuous and quasi-independent, it must have small and isolated effects on gene expression profiles. If it affects the expression of only one gene, that looks like quasi-independent mutation to gene expression. If it affects expression of a gene in only a small proportion of the microenvironments faced by cells, then that looks quasi-independent and continuous at this level too.

I qualify these claims about the appearance of gradualism at the level of gene expression because small changes in developmental mechanisms may make profound differences to the outcome of development. For example, the external cheek pouches used for storing food by gophers and kangaroo rats evolved from internal cheek pouches, which are less adaptive because they lose moisture to the food. The crucial mutation was a rather small change in location and magnitude of epithelial evagination at the corner of the mouth (Brylski and Hall 1988a,b). Ultimately, crucial switches such as this may be due to only a modest change in the expression profile of one gene, so a small change in gene expression may not result in gradual mutation. However, such an amplification of effect is an exception; typically, gradual mutations to gene expression result in no more than gradual mutation to the development of an entire organism.

Explaining the typical non-amplification of development requires investigation of developmental mechanisms above the level of gene regulation. If I successfully explain why gene regulation can evolve gradually, the value of this explanation in terms of the evolvability of organisms rests on acceptance of developmental non-amplification, if not on its explicability. This is a consequence of the reductionist strategy of building from genes up that I outlined at the end of the first chapter. Gene regulation is a low-level aspect of development. Thus, though I am not black-boxing all of development, I am black-boxing much of it. Answers to profound questions about speciation and macroevolution, and perhaps to other questions, may lie in this black box. I suspect that we will find these answers by continuing to build from the gene up while also building from development down.

The theoretical biologist Stuart Kauffman has dedicated much of his career to asking general questions about the adaptivity and the evolvability of gene regulation. He has produced a simple model of gene regulation networks that suggests that natural selection cannot effectively design the regulatory connections between genes. I will present Kauffman's model here, along with my own simple alternative. These will be the simplest models of gene regulation networks presented in this book. Their simplicity allows us to be unusually precise about the evolvability of the model. They might be seen as a case study for modeling gene regulation networks and their evolvability. In the third chapter, I will present Kauffman's more comprehensive model of gene regulation; in the fourth chapter I will present my own model.

2.2 Kauffman's Case Against Natural Selection

In his 1985 article "Self-organization, selective adaptation and its limits," Kauffman argues that natural selection cannot effectively design gene regulation networks. This argument is based on a model that sets selection the task of achieving an optimal network. The success of selection is measured in terms of how closely it can take a population of individuals from randomly generated networks to having optimal networks, despite mutation. Kauffman concludes that natural selection will find it harder and harder to maintain a network as its complexity increases. In view of the complexity of gene regulation networks of even relatively simple life, this poses a serious challenge to the idea that gene regulation networks evolved by natural selection.

To make his case, Kauffman builds a simple model of the evolution of gene regulation networks. The size of the network is determined by the number of nodes (N). Each node is connected to one other node.

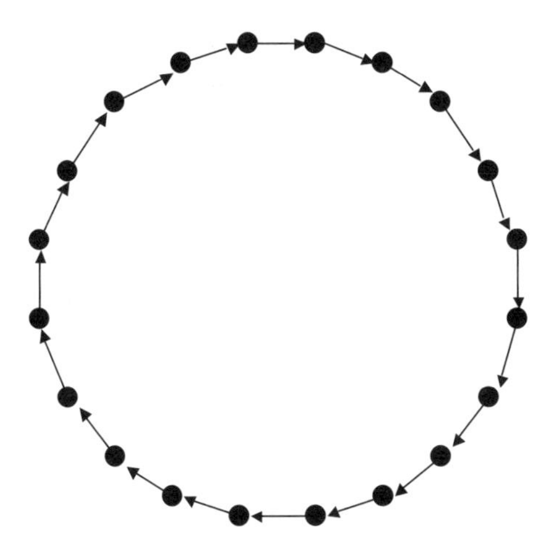

Figure 2.1
An arbitrarily optimal network in Kauffman's simulation.

Kauffman's model assumes that there is one optimal connection for each node. An organism's fitness is determined by how many of these optimal connections it has.

Repeated mutation and selection on a population of networks that were initially produced randomly results in networks with an equilibrium proportion of optimal connections, where the power of natural selection to improve the network is balanced by the power of mutation to degrade it. That proportion depends on the mutation rate, on the fitness advantage that an additional optimal connection grants, and on the number of nodes in the network. The crucial result for Kauffman is that if the mutation rate and the power of selection (i.e., greater fecundity of fitter individuals) are fixed, then an increase in the complexity of the network reduces the proportion of optimal connections that natural selection achieves. If we understand the mutation rate as an absolute number of mutations that happen to a network between every two generations, then an increase in the size of the network makes it harder for natural selection to maintain optimality, because the chance of hitting on the optimal network is reduced (as was explained above). Additionally, if we take the mutation rate as the proportion of connections that mutate (which seems biologically more plausible) and hold that steady with an increase in the size of the network, the increased number of mutations between each two generations will make it even harder for natural selection to maintain an optimal network as the network gets larger.

Whereas increasing the size of the network makes it harder for natural selection to maintain an optimal network, decreasing the mutation rate or increasing the selection coefficient allows natural selection to do a better job. However, the mutation rate must be decreased, or the selection coefficient increased, proportionally to the square of the number of nodes (N^2) to maintain the same level of performance for natural selection. We know that there is significant selective pressure against a high mutation rate because random mutations to effectively designed systems tend to be maladaptive. We also know that evolution has resulted in organisms with ingenious and effective mechanisms to reduce mutation rates. Kauffman's position, however, is that natural selection would be left fighting a losing battle of reducing mutation rates in order to design even more complex genetic networks.

Increasing the fitness advantage of additional optimal connections also increases the level achieved by natural selection at equilibrium. But these fitness advantages have to be very high for selection to be effective in designing complex networks. Kauffman thinks that a model with very high selection coefficients is empirically unrealistic for two reasons. The first is that selection is a noisy process. Although in the majority of cases fitter organisms do better than less fit organisms, bad things often happen to fitter organisms and good things often happen to less fit organisms. Therefore, Kauffman thinks that small differences in the network are unlikely to produce the differences in numbers of progeny that are required to maintain a complex network despite mutation. The second reason is Kauffman's claim that if fitness differences really were that different, then natural selection would not maintain the variety required to face changing environments.

The gene regulation networks of members of most species are highly complex. Kauffman's model suggests that these networks will have a high proportion of maladaptive connections. Therefore, Kauffman's model challenges the view that gene regulation networks have been effectively designed by natural selection. In view of the importance of gene regulation to development, Kauffman's work questions the conventional view that life has been effectively designed by natural selection.

2.3 Modeling Evolvability

Kauffman's model suggests that natural selection cannot achieve a high proportion of optimal gene regulation connections in complex organisms. This indicates that the connections in this model are not sufficiently evolvable. The evolvability of a system is determined in part by the relative fitness of variants of that system. The other aspect of evolvability is how similar highly adaptive variants are to each other. Systems in which highly adaptive variants are similar to each other tend to be more evolvable. The system has one adaptive peak for natural selection to advance toward, with steps toward that peak tending to be more adaptive and therefore favored by natural selection. As it turns out, Kauffman's model and the alternative that I discuss below have one adaptive peak and are therefore more evolvable than they would be otherwise.

Charles Darwin was famously a gradualist, instinctively aware that random changes to a moderately well-designed system have a much better chance of being adaptive if they are slight rather than severe. Richard Lewontin's notions of quasi-independence and continuity (discussed in section 1.7) are both manifestations of gradualism. Quasi-independence allows one trait to change relatively independently of others, thereby reducing the overall change in the ecological relationship. Continuity allows small differences in trait value to make only a small difference in the ecological relationship, too. We can diagnose why Kauffman's model of gene regulation networks cannot be effectively designed by a selection process by showing that its traits are not continuous.

These notions are contingent on the relative fitness of variants. In Kauffman's simulations, the fitness of an organism is determined by the number of optimal connections that it has: an organism with X optimal connections will be half as fit as an organism with $2X$ optimal connections. Each connection is a quasi-independent trait. In fact, they are completely independent, because each connection may mutate independently of the others and because the fitness value of each connection is also determined independently of the others. However, because most changes in trait value are fitness neutral, each connection is not continuous. Only changes that result in the gain or loss of optimality affect fitness, so no mutation can make a connection just slightly better or worse. The degree to which Kauffman's model is gradualistic can be shown graphically by

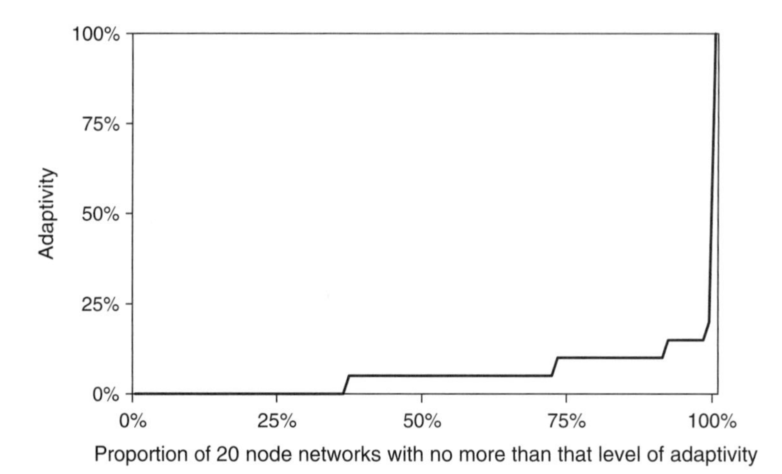

Figure 2.2
The fitness landscape of Kauffman's non-gradualist networks.

means of a fitness landscape. The simplicity of the model makes it easy to determine the fitness of every possible 20-node network with one connection per node, from least fit to most fit (figure 2.2). The least fit 36 percent of possible networks all have no adaptive connections and therefore have zero adaptivity. The next least fit networks have only one adaptive connection. They are 5 percent adaptive, and another 36 percent of networks fall into this category. After that, the proportions get smaller until we reach the only possible network that is 100 percent adaptive. This simple way of constructing a fitness landscape would not show multiple adaptive peaks. But although consideration of multiple peaks is likely to be important when investigating natural systems, we know that there is only one adaptive peak in this model, because the fitness contribution of each connection is determined independently of all other connections.

Horizontal lines on a fitness landscape show regions in which selection is unguided. A lineage is free to drift along the fitness plain without selection's providing any influence. In particular, selection cannot push the lineage toward taking the next step up in fitness. Once such an adaptive mutation happens, selection can work to stop the lineage from falling back down, but until then it has nothing to work with. The abundance of long horizontal plateaus on the fitness landscape of Kauffman's model networks shows their lack of continuity. It shows that selection cannot gradually improve the adaptedness of these networks. This leads naturally to the question "How much better can selection design a model of gene regulation with connections that mutate continuously?"

2.4 Re-assessing Kauffman's Networks

In order to assess how much better natural selection can design a gradualistic model of gene regulation, I developed a continuous, but otherwise analogous, model. In this model, a connection is not simply adaptive or not adaptive; rather, there is an optimally adaptive trait value, with the adaptedness of any connection determined by closeness to that trait value. If the trait value can vary between 1 and 20 and the optimal trait value is 10, a trait value of 9 or 11 is 10 percent less adaptive than a trait value of 10, a trait value of 8 or 12 is 20 percent less adaptive, and so on. The adaptivity of such networks is normally distributed. The fitness landscape of this model is illustrated in figure 2.3.

A generic network has an adaptivity level equal to the median of all possible variants of that network type. It is the expected adaptivity level of a network with no selective history. Therefore, measures of designedness that rate both types of networks fairly will each set a baseline at the

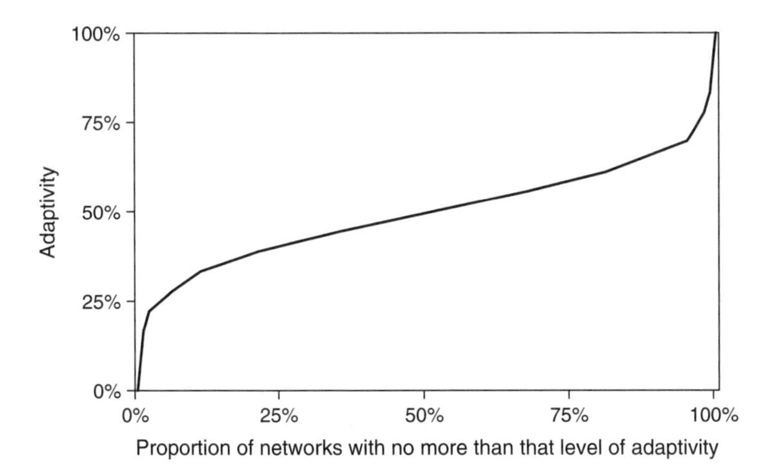

Figure 2.3
The fitness landscape of gradualist networks.

adaptivity of a generic network. By setting this network as a point of 0 percent designedness and the optimal network as 100 percent, I determined the scales of design of the different models, with their different fitness landscapes. I then compared the designedness of each under appropriately similar selection regimes.

In Kauffman's non-gradualist network of 20 connections, the median network has one adaptive connection, whereas an optimal network has 20. The difference provides a linear scale of adaptivity, such that a network with 3 adaptive connections is 10.5 percent designed, for example. In my gradualist network, because the median network is not designed but still is 50 percent adaptive, the scale of designedness sets this value as 0. Therefore, a 50-percent-designed network would have connections that each averaged 2.5 nodes from the optimal value of 10.

The results of simulations on populations of increasing network complexity are illustrated in figure 2.4. In each case, the number of generations is sufficient so that the population is at an equilibrium level of designedness. The results show that as the complexity of the network increases, the population of gradualistic networks is significantly more designed at equilibrium than the population of Kauffman networks.

Kauffman presented a non-gradualistic model of gene regulation; I presented a variant of his model that allowed natural selection to do a better

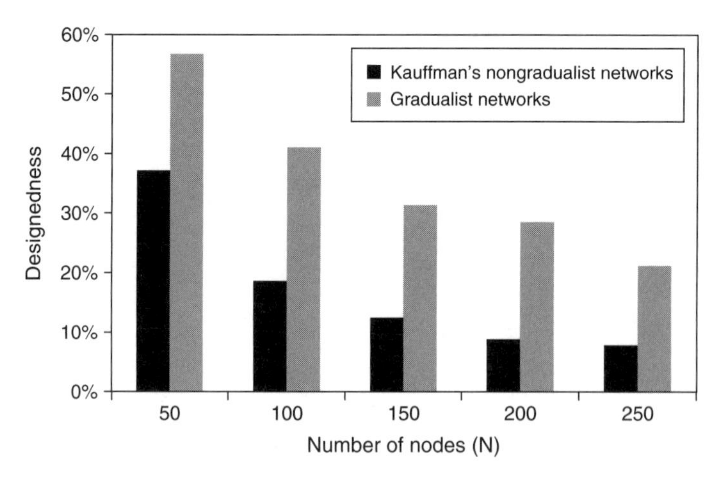

Figure 2.4
Level of designedness achieved by artificial selection.

job. His model was less evolvable than mine. and the design power of selection varied accordingly.[2] Which type of fitness landscape best represents the fitness landscapes of gene regulation networks in organisms? Answering this question definitively requires a great deal of knowledge about gene regulation, development, and ecology. Kauffman's result rested on his assumptions with regard to this point and were not defended empirically.

2.5 Why Model Gene Regulation Networks?

Understanding development from the zygote, and the evolution of development, hence the *evolvability* of ontogeny, requires understanding how such parallel-processing dynamical systems might give rise to an organism, and be molded by mutation and selection. (Kauffman 1990, p. 137)

Despite what his early work suggests, Kauffman has accepted that natural selection is involved in gene regulation. Kauffman developed a more comprehensive model that suggests that the patterns of gene regulation activity are evolvable, and has used it as the basis of a broad framework for understanding how gene regulation networks function in development and how they evolve. I will discuss this model and criticize Kauffman's framework in chapter 3. In chapter 4, I will present my own framework, which is also based on a model of gene regulation networks. That is a lot of models. Just what value is there in any of these models?

Gene regulation networks are a relatively recent focus of research in genetics. Because the relationship between the DNA sequence of regulatory regions and regulatory relationships is complicated, describing the gene regulation network remains challenging, and according to Wray et al. (2003, p. 1402) "no general framework exists for understanding, interpreting, and predicting how transcription evolves." Stuart Kauffman and I differ in our views of gene regulation and development, but we both believe that insight into gene regulation can lead to insight into development and its evolution. There are a number of steps between modeling gene regulation networks and explaining the adaptivity and evolvability of organisms. Models are abstract simplifications of reality—misdescriptions, if you will. Science often progresses through the use of models, but exactly how such progression works, or should work, is not straightforward. (See, e.g., Black 1962; Hempel 1965; Suppe 1989.) Good models are, in a way, the right misdescriptions, because they allow us to focus on the most crucial and explanatory components of reality. Kauffman's model assumed that the adaptedness of potential regulatory connections in gene regulation networks was either adaptive or maladaptive (with a significant majority maladaptive); my model assumed that they varied on a continuum. Neither assumption was defended empirically, but the difference in the evolvability of these models indicates that whether the adaptivity of gene regulatory connections varies discretely or whether it varies continuously may be crucial to our understanding of the evolution of gene regulation networks. Thus, Kauffman's model and my model indicate that empirical investigation of the distribution of adaptivity of gene regulatory relationships is important.

Even if a good model of gene regulation networks were to be developed, one might still question what could be concluded about development and its evolution from that model. It is difficult to produce an adequate argument showing that, in order for development to be a certain way, it is necessary that gene regulation be a certain way, or showing that a certain type of gene regulation is sufficient for certain type of development, because development can produce emergent phenomena at levels above gene regulation. For example, the density of our bones is determined in part by the stresses put on those bones during development. This is adaptive developmental variation with no variation in genes or their regulation. However, owing to the ubiquitous influence of gene regulation on phenotype (Wray et al. 2003), our view of gene regulation loses plausibility if it is completely at odds with our view of development. For example, if we were to find that the assumptions of Kauffman's model were empirically justified, and that natural selection could not design the regulatory relationships between genes, that would have serious implications for what we should expect development to be. In view of the importance of gene regulation to development, either the connections themselves are not particularly important but some emergent property of gene regulation is important (as Kauffman believes) or the connections themselves are important and gene regulation is a source of random noise in development. If the latter is the case, then development could not be the result of evolution by natural selection and either is designed by something else or is not the intricate and superb example of engineering that we take it to be.

If any conclusion from the nature of gene regulation to the nature of development would be reasonable, then investigating gene regulation networks by use of models is relevant to how we should understand the development and the evolution of organisms. Kauffman and I hold different views of gene regulation that recommend different views of development and evolution, but we both expect evidence about one to be relevant to conclusions about the other.

3 Kauffman's Framework for Gene Regulation

Stuart Kauffman's more comprehensive model of gene regulation is the result of more than thirty years of research. His two books on the subject, *The Origins of Order* (1993) and *At Home in the Universe* (1995), represent the most developed theoretical approach to gene regulation networks. Kauffman's goal in investigating gene regulation networks is different from mine. He is primarily interested in explaining the complex order of development and of gene regulation; I am primarily interested in explaining the evolvability of development and of gene regulation.

Organisms are low-entropy states. The second law of thermodynamics tells us low-entropy states are out of equilibrium. Darwin's theory of evolution by natural selection offers a general explanation for why organisms exist far from thermodynamic equilibrium and maintain a low-entropy state over their lifetime. Natural selection is not a force that holds anything together, but it explains the widespread presence on Earth of things that—owing to the laws of physics, chemistry, and so on—just happen to use energy to hold themselves together. These things exist because their ancestors were able to reproduce.

Kauffman has expressed dissatisfaction with the standard Darwinian explanation for the order of organisms because it leaves complex organisms (such as humans) "accidental." Instead, Kauffman seeks a view that shows that we are "expected" (1995, p. 8):

> Random variation, selection sifting. Here is the core, the root. Here lies the brooding sense of accident, of historical contingency, of design by elimination. At least physics, cold in its calculus, implied a deep order, an inevitability. Biology has come to seem a science of the accidental, the ad hoc, and we are just one of the fruits of that ad hocery. Were the tape played over, we like to say, the forms of organisms would surely differ dramatically. We humans, a trumped-up,

tricked-out, horn-blowing, self-important presence on the globe, need never have occurred.

Where, then, does this order come from, this teeming life I see from my window: the urgent spider making her living with her pre-nylon web, coyote crafty across the ridgetop, muddy Rio Grande aswarm with no-see-ems (an invisible insect peculiar to early evenings)? Since Darwin, we turn to a single, singular force, natural selection, which we might as well capitalize as though it were a deity. Random variation, selection-sifting. Without it, we reason, there would be nothing but incoherent disorder.

I shall argue in this book that this idea is wrong. For as we shall see, the emerging sciences of complexity begin to suggest that the order is not at all accidental, that vast veins of spontaneous order lie at hand. Laws of spontaneous order generate much of the order of the natural world. It is only then that natural selection comes into play, further molding and refining. (1995, pp. 7–8)

Kauffman attempts to explain the order of organisms by studying the order of gene regulation networks, and he attempts to explain the order of gene regulation networks by studying the order of generic networks. A generic network is the average network of many networks produced without any design. Its features are due entirely to its general architectural structure. Thus, Kauffman attempts to explain the order of organisms by studying the order of generic networks. In doing so, he hopes to show that we are not the results of the chancy business of mutation and selection, but are instead the "expected" results of generic order. This offers us a naturalistic analog for the security that we had under theistic views of the world, which place us at the center of importance owing to an act of divine will. Accordingly, we can then feel "at home in the universe."[1]

3.1 Kauffman's Model of Gene Regulation

Kauffman builds his case for the generic order of gene regulation networks with models. In these models, each gene is represented by a node that is "on" or "off." Whether a node is on or off at time t_n is determined by the activity of its input nodes at t_{n-1}. The connectivity (K) of a network represents how many input nodes each node has. The activity that would be produced by a node at t_n after each possible combination of activity of its inputs at t_{n-1} makes up the function of that node. In a Kauffman network, node function and the identity of input nodes are assigned randomly to each node at the creation of the network.

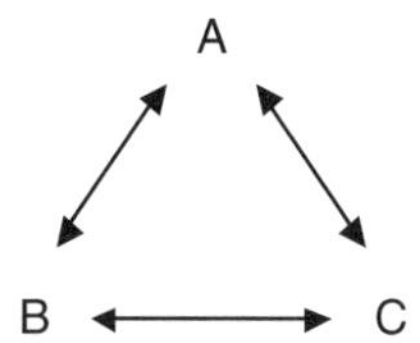

at t_n		at t_{n+1}
B and C		A
0	0	0
1	0	0
0	1	0
1	1	1

at t_n		at t_{n+1}
A and C		B
0	0	0
1	0	1
0	1	1
1	1	1

at t_n		at t_{n+1}
A and B		C
0	0	0
1	0	0
0	1	1
1	1	1

Figure 3.1
A simple Kauffman network (Sansom 2008a).

Figure 3.1 shows a very simple Kauffman network. It is made up of only three nodes, and each of them has the other two as inputs (i.e., $K = 2$). For example, the activities of nodes A and B at t_n determine the activity of node C at t_{n+1}. The function of each node is also shown, with 1 representing that a node is "on" and 0 representing that a node is "off."

Kauffman's models of gene regulation networks are discrete, Boolean, and deterministic. In all three respects, these models are simplifications of reality.

First, Kauffman's models are discrete, because they divide time into discrete parts. States at one moment cause states at another. In the real world, we expect time to be continuous (i.e., within any period of time, there are shorter periods of time). Thus, Kauffman's divisions of time are unrealistic. If time were discrete, the smallest real period of time would be much shorter than the periods represented in the model. Finding that gene expression is due to the presence of other genes, and that it takes between 1 minute and 10 minutes to start or stop the expression of a gene, Kauffman assumes that the presence of genes at one moment will determine the expression of genes at a moment between 1 and 10 minutes later, which will, in turn, determine the expression of genes at a moment between 1 and 10 minutes still later. Practically speaking, the model divides time into periods representing 1 to 10 minutes but ignores fluctuations within that period. For example, whether a particular gene is expressed more in the first or last seconds of a period is ignored.

Second, Kauffman's models are Boolean, because they represent a gene's activity over a short period of time as being in one of only two states ("on" and "off") and ignore the fact that gene expression levels vary beyond simply being expressed or not. Over a 1–10-minute period in which genes X and Y are expressed, more proteins might be produced from gene X than gene Y. Thus, Kauffman's model ignores the variety of levels of expression that can occur in the "on" state. Practically speaking, Kauffman's model requires that a gene be expressed at a relatively low level for it to count as "on" and treats all genes expressed at this level the same way.

Finally, in Kauffman's models, the state of gene activity a_1 at time t_1 causes the state of gene activity a_2 at time t_2 deterministically. This is deterministic because a_1 at t_1 has 100 percent probability of causing a_2 at t_2, rather than having, for example, an 80 percent chance causing a_2 at t_2 and a 20 percent chance of causing a_3 at t_2. The mechanisms of gene expression are almost certainly genuinely indeterministic, because they involve molecules attaching to each other, where the movement of electrons plays a role, which the most widely accepted view of quantum mechanics has shown to be indeterministic. So, for example, if the chance of a specific transcription factor binding to a gene's regulatory region within a short range of time is 90 percent, whether or not it actually does bind is thought to be genuinely indeterministic and therefore, in principle, unpredictable. Some have denied the indeterministic interpretation of quantum mechanics, and some may deny that these indeterminacies affect phenomena at the level of protein bonding. Regardless, we will never be able to predict the specific details of exactly when any protein will be produced in an organism, because there are too many factors involved.

How can we justify using a deterministic model for an indeterministic system? In general, the justification for deterministic predictions for indeterministic systems rests on the law of large numbers. Assume that the tossing of a fair coin is indeterministic. We cannot accurately predict the result of any one toss, but we can predict with a high degree of accuracy and certainty how many times the coin will come up "heads" in a million tosses. Gene expression due to transcription factors' binding, then unbinding, then binding again many times in a 1–10-minute period is assumed to be highly predictable in the same way.

The assumption behind Kauffman's modeling of gene regulation networks as discrete, Boolean, deterministic systems is that each simplifying assumption is supported by the other two. Whether a gene is expressed at a high enough level to count, during a 1–10-minute period, is assumed to be highly predictable, in principle.

Given enough time, every Kauffman network will fall into a repeating pattern of activity. Because there are only so many possible states of a Boolean network of any size, eventually it must hit on the same state a second time. Because it is deterministic, the network will then take the same path to return to that same state. These repeating patterns of activity are called "attractors." The network illustrated in figure 3.1 has three attractors. Two of these repeat the same state over and over; the third repeats a pair of states. The attractors are illustrated in figure 3.2. No matter in which state the network starts, it will end up in one of these attractors. It takes no more than two state sequences for the network to enter one of the attractors. Figure 3.3 shows which initial states result in which attractors. The set of states that lead to a particular attractor are called its "basin." In this network, attractor 3 has the largest basin, with four states.

All deterministic Boolean networks have attractors, but these repeating patterns of activity can be so long that the order is effectively lost. If an attractor has so many steps that it cannot be repeated by a cell before it

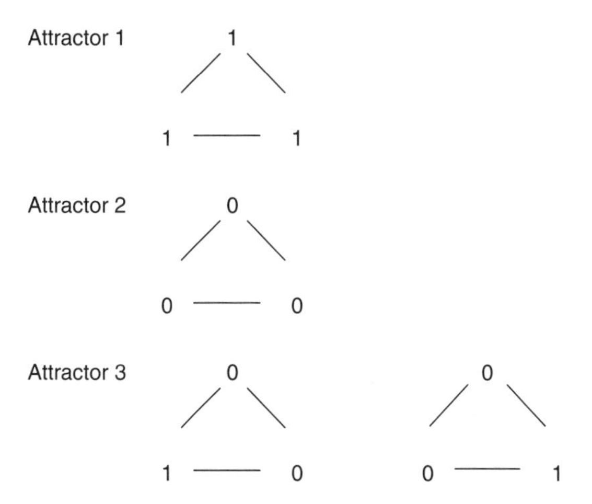

Figure 3.2
The attractors of a simple Kauffman network (Sansom 2008a).

A	B	C	Attractor
0	0	0	2
0	0	1	3
0	1	0	3
0	1	1	1
1	0	0	3
1	0	1	3
1	1	0	1
1	1	1	1

Figure 3.3
The basins of the attractors of a simple Kauffman network (Sansom 2008a).

dies, then it is no example of order. The length of the median attractor in a generic Kauffman network is determined by three factors. The first two are the number of nodes (*N*) and the connectivity (*K*). *K* is the number of input nodes for each node. Increasing either *N* or *K* increases attractor length. Consider a network with 200 nodes ($N = 200$), and with each node determined by the activity of two nodes ($K = 2$). The median length of the attractors of this network is repeating patterns of activity of just 14 states. But if connectivity is increased to $K = 200$ in Kauffman's networks, the median attractor increases in length to 10^{30} states. If the network took a millionth of a second to progress from one state to another, it would still require billions of times the history of the Earth (4.6 billion years) to get through one complete cycle of an attractor (Kauffman 1995, p. 82). And this is only a network of $N = 200$. Given general physical limitations, generic Kauffman networks of any size with high values of *K* effectively show no order at all.

In contrast, Kauffman networks with low values of *K*, particularly $K = 2$, show great order. The length of attractors of $K = 2$ networks are typically approximately the square root of *N* (Kauffman 1995, p. 83). Even a network with $N = 100{,}000$ has attractors of only 317 states.[2] A network need only satisfy this general architectural requirement to be orderly; it need not be precisely designed to have order by selection. According to Kauffman's model, each network attractor represents a cell type. That is to say, Kauffman's model assumes that the activity of genes determines the activity of a cell, so for a cell to be of a particular type is for the genes in a cell to be repeating a pattern of gene expression, determining that the cell repeat a particular pattern of activity. If perturbation causes a cell to jump from one repeated cycle of activity to another, that will change its cell type. Thus, Kauffman thinks that attractors are the key to explaining the order of cellular activity, including cellular differentiation and morphogenesis. Kauffman finds the short, orderly attractors that he thinks are crucial to orderly development in his randomly generated networks of $K = 2$. This suggests to Kauffman that $K = 2$ in gene regulation networks.

3.2 We the Expected?

Below, I shall offer a number of criticisms of Kauffman's explanation of the order of organisms. Here, I want to question whether Kauffman's proposal that $K = 2$ gene regulation networks have generic order is necessary or sufficient for achieving his expressed goal of showing that we can feel "at home in the universe" because we are "expected" rather than "accidental."

Stephen Jay Gould viewed the course that evolution by natural selection has taken on this planet as highly contingent: "Replay the tape a million times from a Burgess beginning, and I doubt that anything like *Homo sapiens* would ever evolve again. It is, indeed, a wonderful life." (1989, p. 289) Others have argued that evolution by natural selection is less contingent. For example, Simon Conway Morris (2003) has argued that the abundance of evolutionary convergence shows us that, once evolution gets underway, the evolution of something like *Homo sapiens* is almost inevitable. Conway thinks that this is because natural selection consistently favors the same solutions to life's problems. The quotation at the beginning of this chapter expresses Kauffman's dissatisfaction with Gould's view of the contingency of human evolution. However, a proposed solution such as Morris's appears insufficient for Kauffman too, because it proposes that natural selection will inevitably hit upon something human-like, and Kauffman expresses dissatisfaction with the "random variation, selection sifting" that is natural selection itself—no matter how inevitable its results, presumably.

Gould's position on the contingency of life was not completely straightforward. Plainly he thought that much about life is not contingent at all:

I suspect that the origin of life on earth was virtually inevitable, given the chemical composition of the early oceans and atmospheres and the physical principles of self-organizing systems. Much about the basic forms of multicellular organisms must be constrained by the rules of construction and good design. The laws of surfaces and volumes, first recognized by Galileo, require that large organisms evolve different shapes from smaller relatives in order to maintain the same relative surface area. Similarly, bilateral symmetry can be expected in mobile organisms built by cellular division. (1989, p. 289)

Gould explicitly limited the prominent role that he gave to contingency in evolution to matters of detail:

. . . these phenomena, rich and extensive as they are, lie too far from the details that interest us about life's history. Invariant laws of nature impact the general forms and functions of organisms; they set the channels in which organic design must evolve. But the channels are so broad relative to the details that fascinate us! The physical channels do not specify arthropods, annelids, mollusks, and vertebrates, but, at most, bilaterally symmetrical organisms based on repeated parts. (ibid., pp. 289–290)

The detail that Gould repeatedly described is the evolution of *Homo sapiens*:

Why did mammals evolve among vertebrates? Why did primates take to the trees? Why did the tiny twig that produced *Homo sapiens* arise and survive in Africa? When we set our focus upon the level of detail that regulates most common questions about the history of life, contingency dominates and the predictability of general form recedes to an irrelevant background." (ibid., p. 290)

However, the breadth of these contingent details varies across passages. Gould concluded his book with other contingencies in the evolution of *Homo sapiens* that have broad implications. He argued that dinosaurs would have continued to occupy all large-body ecological niches had it not been for a mass extinction due to extraterrestrial impact. Even more profoundly, chordates may not have evolved if *Pikaia* (a two-inch ribbon-shaped organism and the earliest known chordate)—a creature with no apparent extraordinary adaptivity—hadn't been lucky enough to survive the Burgess decimation.

Kauffman's dissatisfaction with evolutionary explanations lies in the fact that they leave "us" improbable. Just as we need to know how broadly Gould construed the breadth of the domain for contingency in evolution, we need to know how specific the expectability Kauffman advocates is. Is this a modest project of replacing the improbability of any complex species with expectability, or is it an ambitious project of showing the expectability of a something much like *Homo sapiens*? In the quotation at the beginning of this chapter, Kauffman explicitly mentions human beings. But elsewhere in the same book (*At Home in the Universe*) he refers to organisms in general:

> In this view of the history of life, organisms are cobbled-together contraptions wrought by selection, the silent and opportunistic tinkerer. Science has left us as unaccountably improbable accidents against the cold, immense backdrop of space and time. (1995, p. vii)

The ambiguity is crucial, because, even if Kauffman's arguments about the generic order of life are successful, he achieves only the modest goal of making complex organisms probable—not *Homo sapiens* in particular, or anything resembling a human. Some passages in *At Home in the Universe* suggest that Kauffman accepts as much. For example:

> For the hope arises that viewed at the most general level, living systems—cells, organisms, economies, societies—may all exhibit lawlike properties, yet be graced with a lacework of historical filigree, whose wonderful details that could have easily been otherwise, whose very unlikelihood elicits our awed admiration. (ibid., p. 19)

In that case, Kauffman's position on the role of contingency may not be as radical as he implies when he contrasts himself with Gould. Kauffman may only be elaborating the predictability of general form that Gould considered the irrelevant background, after all.

Given that there really does not appear to be an argument for the ambitious project, let us accept that the relevant issue is whether the presence of some type of complex organisms should be expected, rather than something human-like. Kauffman offers an explanation of why complex organisms are highly probable. What about Gould? In his 1997 book *Full House: The Spread of Excellence from Plato to Darwin*, he championed the idea that an increase in the complexity (or excellence) of the most complex species on Earth can be explained as the result of a random walk from a point of origin of low complexity. It should be noted that Gould did not explain how mutations can take organisms from simple to complex, which this book concerns itself with. Gould simply assumed that mutations that add complexity and mutations that take it away occur regularly, so a random walk from a minimal average complexity starting point will take the population of organisms of the Earth to higher average complexity.

This type of explanation in evolution that invokes a diffusion-with-boundary principle was first developed by Steven Stanley (1973) and

Dan McShea (1994) was first to apply it to the evolution of complexity. McShea (1996) distinguished four types of complexity and looked for measures to discover whether we can show that there really is a bias in evolution toward complexity. McShea's survey found progress in metazoans as a whole in only one type of complexity, and only during the early Phanerozoic. Trends in other types of complexity were limited to specific metazoan subgroups. Taken as a whole, this evidence suggests that life has generally not been driven to greater complexity, but has on average gotten more complex because it started at a point where it was almost as simple as it could be.

Gould's favored diffusion-with-boundary explanation leaves the evolution of complex organisms from the simple origin of life to be expected largely as a matter of logic. Kauffman's laws of complexity about the generic order of certain sorts of systems that are supposed to make complex life expected are ultimately a matter of logic as well. Therefore, Kauffman's arguments may not leave complex life any more expected than other, more standard interpretations of evolutionary theory.

Putting aside the issue of our contingency, Kauffman still presents a curious and radical view of life and evolution that is worthy of consideration. The goal of the present book is to contribute to explaining the evolvability of development by explaining the evolvability of gene regulation networks. Kauffman's goal is to contribute to explaining the order of development by explaining the order of gene regulation networks. Kauffman's project has relevance to mine because evolvability is a type of order. Generally speaking, evolvability is due to mutations' tending to have smaller effects (thereby giving them a higher probability of being adaptive), which is one type of order. Kauffman focuses his attention on attractors, and measures order in terms of attractor length. Though this is not a measure of evolvability, it is associated with one on his view. Kauffman says that his $K = 2$ networks that have short attractors "can undergo a mutation that alters wiring or logic without veering into randomness." He continues: "Most small mutations cause our hoped-for small, graceful alteration in the behavior of the network. Basins and attractors change only slightly. Such systems evolve readily. So selection does not have to struggle to achieve evolvability." (1995, pp. 83–84)

Kauffman's view implies the following answer to the problem of evolvability: Organisms are evolvable because the activity of development is primarily the repetition of activity by cells, which is regulated by gene regulation networks. The activity of these networks is evolvable because each gene is typically regulated by two other genes, and mutations to the regulatory relationships in such networks result in only gradual changes in the repeated cycles of activity that they regulate.

3.3 Explaining the Order of Kauffman's $K = 2$ Networks

Although Kauffman typically explains the order of his $K = 2$ networks in terms of their low connectivity, he acknowledges another general determinant of order: the type of node functions. In this section, I will argue that Kauffman's $K = 2$ networks are orderly only because they tend to have orderly node functions.

The list of outputs produced by each possible input is the function of the node. There are four possible inputs to a node in a $K = 2$ function (see figure 3.1), and each input can produce an output of 1 or 0, so there are 16 possible functions. In a Kauffman network, each node is randomly assigned one of these functions, so the probability that a node will have the exclusive-or Boolean function (XOR) is 1/16. If functions are assigned differently, such that all nodes are assigned the exclusive-or function, attractors are lengthened enormously. They are longer, in fact, than completely random networks, in which the network state that each network state generates is picked at random. The same is true of networks with the XNOR function or a combination of nodes with either function. To show this, I generated networks of $K = 2$ with all nodes assigned the exclusive-or function. For each network, I produced a random initial state and measured the length of the attractor that it reached (i.e., first attractor length). The median lengths of first attractors of 500 networks for several values of N are shown in figure 3.4. They are compared with the median first attractors of completely random networks and with regular Kauffman $K = 2$ networks that have their functions generated randomly.

Before analyzing Kauffman's model of gene regulation, I wish to explain my rationale for using median first attractor length as my measure of attractor order. The distribution of attractor lengths is strongly positively skewed. For this reason, Kauffman adopts the convention of measuring in terms of the median (as does Luc Raeymaekers, discussed below). This

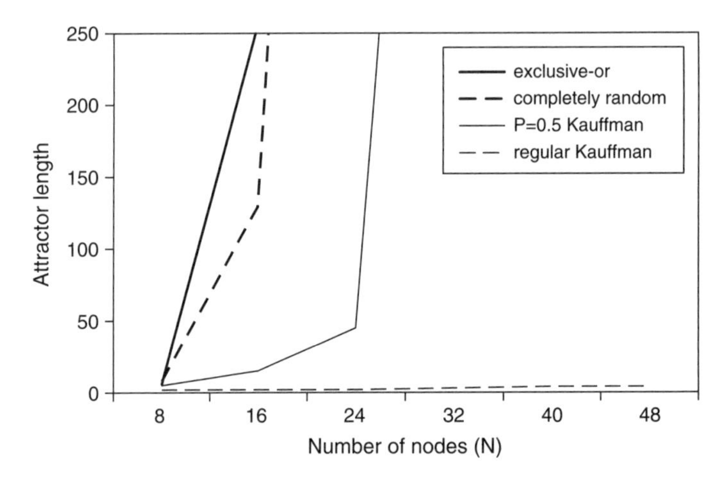

Figure 3.4
Median first attractor length; of 500 exclusive-or networks, of 500 networks in which one state randomly determines the next, of 500 networks with $K = 2$ and $P = 0.5$, and of 500 regular $K = 2$ Kauffman networks (i.e. $K = 2$ networks with randomly assigned functions) (Sansom 2008a).

convention gives a more orderly result, because the median attractor is shorter than the average. Which model measure is most enlightening? If each cell type is locked in a different attractor, then some should be locked into attractors long enough to exhibit no order at all. In addition, longer attractors have more entry points and therefore larger basins, so, if we pick one of a network's short attractors and one of its long attractors, a random initial state is generically more likely to end up in the long attractor than in the short one. With all these issues in mind, I compromised and adopted the measure of median first attractor length. A network's first attractor length is more likely to be a longer attractor of the network than a shorter attractor, because longer networks have larger basins. I used the median first attractor length of different networks to try to adjust for the positively skewed distribution of all attractors.

Exclusive-or networks are highly disordered $K = 2$ networks that show the insufficiency of $K = 2$ to produce orderly networks. These results were produced by manipulating $K = 2$ node functions, which Kauffman acknowledges to be the third source of order beyond N and K.

Kauffman points to two features of these functions that contribute to order. The first is a bias of a node function toward one output (P). For node

A in figure 3.1, $P = 3/4 = 0.75$, because three of the four input activity states of B and C will produce the output 0 for A. Because 0 is a more common output for this node, P is the proportion of input activities that will produce that output. For node B, $P = 0.75$, although in that case 1 is the more common output, so P is the proportion of input activity states of A and C that produce that output. For node C, $P = 0.5$, the lowest possible value. Networks that have nodes with higher average values of P have shorter attractors. The exclusive-or networks discussed above have $P = 0.5$, which partially explains their disorderliness. In contrast, the average value of P in a regular $K = 2$ network is 0.6875.

To show the significance of variation of P in $K = 2$ Kauffman networks, I generated networks of $K = 2$ with a random selection of functions for each node with $P = 0.5$. The median lengths of first attractors of 500 networks for several values of N are also shown in figure 3.4. They show that, though reducing P decreases order, the reduction is not sufficient to explain the disorder of exclusive-or networks.

The second function feature that Kauffman acknowledges to have an influence on attractor length is the presence of canalizing functions, which he defines as "any Boolean function having the property that it has at least one input having at least one value (1 or 0) which suffices to guarantee that the regulated element assumes a specific value (1 or 0)" (1993, p. 203). For example, in figure 3.1, node A is regulated by nodes B and C in such a way that if node B has value 0 at t_n, then node A will have value 0 at t_{n+1} regardless of the value of C at t_n. In this case, the canalizing value of B is the same as the canalized value of A, but that is not necessary for a canalizing function. The canalized value of A could have been 1. All $K = 2$ functions are canalizing except exclusive-or and exclusive-nor, which explains why exclusive-or networks are less orderly than networks of any functions with $P = 0.5$.

The above results suggest to me that the order of Kauffman's $K = 2$ networks is not due to their low connectivity, but rather to a bias in randomly produced node functions of networks with few inputs toward canalizing functions and functions with high P. Therefore, I suggest that we largely leave $K = 2$ networks behind and concentrate on P and canalizing functions. I will argue in the remainder of this section that they are not quite sufficient to fully explain the order that Kauffman is looking for by comparing higher K networks with networks having the same P and the

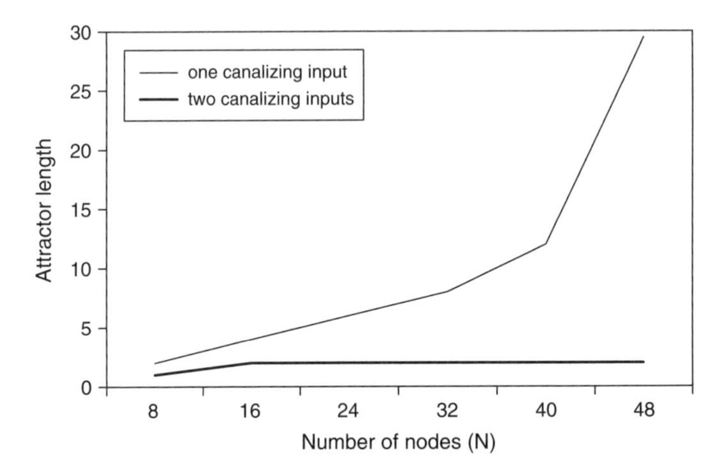

Figure 3.5
Median first attractor of 500 $K = 6$ networks with one canalizing input value and $P = 0.75$ functions and of 500 $K = 6$ networks with two canalizing input values and $P = 0.75$ functions (Sansom 2008a).

same proportion of canalizing functions but different numbers of canalizing input values.

As described above, a canalizing function is one in which at least one value of one input is sufficient to determine the output. That is to say, a canalizing function has at least one canalizing input value. Some canalizing functions have more than one canalizing input value. Biasing a network toward canalizing functions with a higher number of canalizing input values need not result in an increase in P. The networks above have 100 percent canalizing functions and $P = 0.75$. But whereas all functions in the first group have only one canalizing input value, those in the second have two canalizing input values. The second group has significantly shorter attractors, and is therefore more orderly, despite having the same N, K, P, and proportion of canalizing functions. The attractors of the second group are similar to Kauffman's $K = 2$ networks. (See figure 3.4.)

These results discussed in this section probably should be considered an addendum to Kauffman's explanations of the order of his networks. They also act as preliminary results as we move from analyzing the order of Kauffman's networks to judging these networks as models for gene regulation networks. In the remainder of this chapter, I shall consider the biological plausibility of Kauffman's explanations of the order of gene regulation and development, and offer various suggestions regarding the details of Kauffman's explanation, before making a broad criticism and suggesting an entirely different view.

3.4 The Relevance of Kauffman's Explanations of the Order of Gene Regulation

Kauffman makes three empirical claims about gene regulation that are supposed to show the relevance of his model and explanations of the order of $K = 2$ networks to the order of gene regulation networks (modified from Kauffman 1995, p. 103). These are the most important tests that scientific models face, because this is where the model confronts reality.

The first of Kauffman's claims is that *the Boolean functions of expression differ across genes*. Functions are determined randomly in Kauffman's networks. Kauffman suggests that the variation in the expression functions of genes shows the applicability of this assumption in the model. Though it is true that node functions and gene expression functions vary, I will argue in the next section that the random functions in Kauffman's models are unrealistic and have resulted in the loss of an important source of order in gene regulation networks.

The second claim is that *genes are directly regulated by rather few molecular inputs*. Kauffman expects this to be true in part because such networks are chemically easy to build. However, the chemical complexity of DNA appears to have resulted in significantly more than two transcription factors' influencing at least some genes.

It has proved hard to conclusively discover an approximation for the average number of transcription factors that directly regulate genes in complex organisms. A number of different techniques are used, and they indicate different answers. One technique involves observing the changes is the presence of proteins in cells over time. Statistical tools are used to discover the correlations between the levels of different proteins over time. This allows the regulatory relationships to be induced with appropriate statistical certainty. Such methods have led to estimates that the number of transcription factors regulating each gene is fairly low. Averaging across studies of a large number of genes in the bacterium *Escherichia coli* suggests $K \approx 1.4$ and $K \approx 1.7$ in the budding yeast *Saccharomyces cerevisiae* (Shen-Orr et al. 2002, Rosenfeld and Alon 2003; Rosenfeld and Alon 2003; Costanzo et al. 2001; Lee et al. 2002). In more complex organisms, complexity looks higher: $K \approx 1.9$ in the fruit fly *Drosophila melanogaster* and $K = 2.8$ in *Arabidopsis thaliana* (a small flowering plant) (Serov et al. 1998, Rosenfeld and Alon 2003). But such methods probably underestimate regulatory connections, because it shows only the effects of transcription factors that are present during the study. For example, if gene X's regulatory region has a binding site for activating transcription factor Y, but Y is never present during the study, Y will not be recognized. This is particularly important in the study of complex organisms because different transcription factors can play the crucial role in regulating the same gene in different circumstances (e.g., in different cell types). This effect will be mitigated to some degree in studies of the simplest organisms, such as *E. coli* and *S. cerevisiae*, but in complex organisms the question becomes What proportion of regulatory connections do these methods miss?

Other methods involve more detailed experimental work, such as identifying potential regulatory connections, producing mutations to disrupt them and seeing if it has an effect. Some eukaryotic regulatory regions now are understood quite well. (See, e.g., Arnone and Davidson 1997; Wilkins 2002.) This work has led to the estimate that 10–50 binding sites of 5–15 different transcription factors is not unusual (Wray et al. 2003). Kauffman's networks of $K > 3$ lose the order of attractors that he used to explain the order of gene regulation, unless they have a high proportion of sufficiently canalizing functions or functions with sufficiently high values of P. I agree with Wray that, on balance, the limited empirical evidence favors the hypothesis that genes are regulated by a relatively high number of transcription factors, certainly more than the number proposed by Kauffman.

The third of Kauffman's claims is that *genes are regulated by canalizing functions (at least in the cases we know)*. High-K Kauffman networks can maintain some degree of order if they have a high proportion of sufficiently canalizing functions, so any denial of Kauffman's second claim makes the truth of the third crucial for the applicability of his model. My own analysis above suggests that low K is orderly predominantly because such networks have a high proportion of canalizing functions, anyway. So Kauffman's case relies hugely on this empirical prediction. Results discussed above (see figure 3.5) suggest that canalizing functions with only one canalizing input value will not be sufficient to provide the order

Kauffman is looking for, although canalizing functions with multiple canalizing input values might be.

The empirical evidence that Kauffman's view requires is not of the sort typically sought out by molecular biologists, having nothing to do with adaptive gene expression and not being readily available in the literature. Until empirical evidence of the required kind is accumulated, all we can say is that whether enough genes are regulated by functions with enough canalizing input values remains an unanswered question that must turn out in Kauffman's favor if his model is to be applicable to gene regulation.

3.5 Additional Orderly Facts of Transcription

Some factors not incorporated in Kauffman's model of gene regulation increase the orderliness of gene regulation networks by Kauffman's measure (i.e., reduce attractor length). Kauffman's model has every gene as likely as any other to influence the transcription of another gene. But in fact only a small proportion of genes produce transcription factors, which are capable of binding to DNA on the regulatory region of specific genes and, thereby, directly influencing its transcription rate. The rest have completely different functions. At the upper end of the spectrum of genetic complexity, for example, we have between 20,000 and 25,000 genes (International Human Genome Sequencing Consortium 2004), but only about 1,850 of them are thought to produce transcription factors (Venter et al. 2001). Kauffman was concerned with the attractors in a network of all genes. Given that the transcription rates of all genes are due to the presence of transcription factors, I suggest that it is more appropriate to model a network of transcription factors only (transcription factors regulate each other, too), which I call *the transcription factor regulation network*. Be that as it may, significantly reducing the size of the applicable Kauffman network still leaves them well in the disorderly realm if K remains high and canalizing functions are not abundant. For example, Kauffman networks of $N = 1{,}850$ with K values of only 4 or 5 still show "similar" disorder to networks of $K = 1{,}850$, which has expected cycle lengths of 2,925 states (Kauffman 1995, p. 82).

There is order in the connections of gene regulation networks that Kauffman's model does not take into account. Empirical evidence suggests that the influence of each transcription factor on each of its target genes is nearly always *qualitatively consistent*. That is to say, all transcription factors either "activate" the transcription of a target gene or "repress" it. As I will explain below, this is related to but not identical to the concept of a "linearly separable" function, in which the influence of each node is cumulative in a way that is independent of the activity of other inputs. I call linearly separable functions (as in figure 3.6) *quantitatively consistent* to distinguish them from qualitatively consistent functions (as in figure 3.7).

The table in figure 3.6 represents a quantitatively consistent function (the function B). All possible activity combinations of A and B are shown in the first two columns and the output they produce (i.e., the activity of C) is shown in the third. A function is quantitatively consistent if it can be achieved by a two-layer quantitatively consistent (i.e., linearly separable) network. Such a network has only input and output nodes, and each input's additive contribution to the output is independent of the activity of other inputs. Figure 3.6 is a simple example of such a network with connection values shown. In this Boolean network, each node follows the rule that the activity of a node will be 1 if that sum of its inputs is greater than 0, and will be 0 otherwise.

The quantitative consistency of a function is sufficient but not necessary for its qualitative consistency. In a qualitatively consistent function, no input activates the output in one context and represses it in another. Consider the third and fourth rows of the function table in figure 3.6. A is active in both rows, so the rows represent the context of A activity. B effectively activates in the context of A activity, because the output is

	Input A activity	Input B activity	Output C activity
1	0	0	0
2	0	1	1
3	1	0	0
4	1	1	1

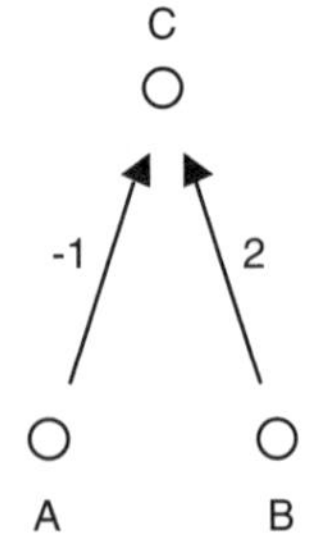

Figure 3.6
Quantitatively consistent node function of output C by inputs A and B (Sansom 2008a).

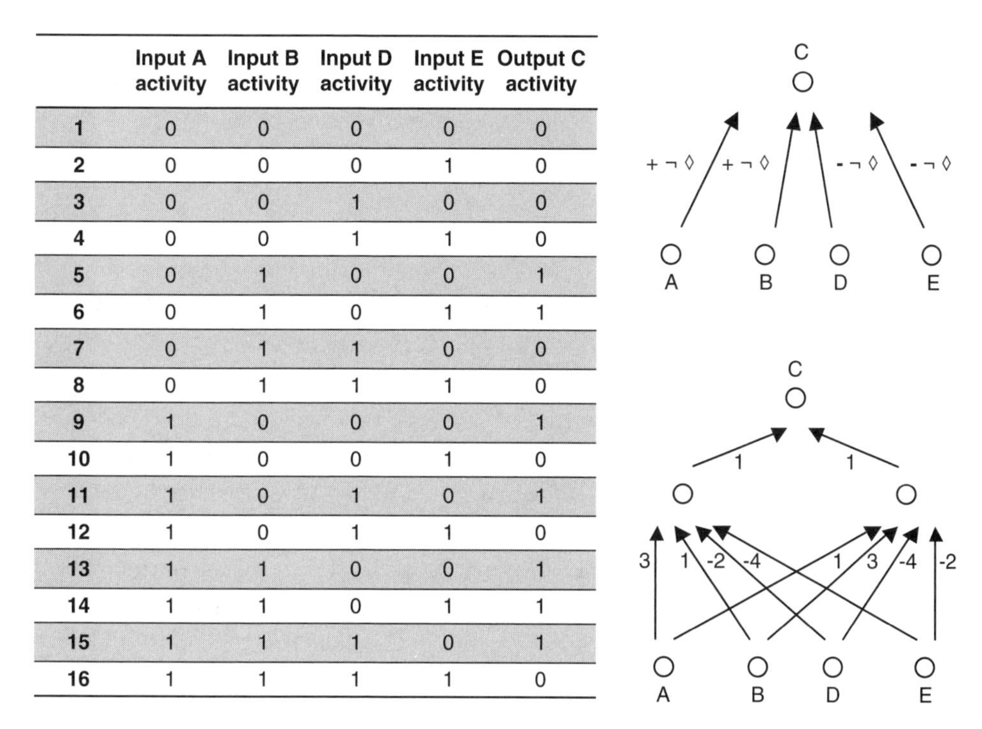

	Input A activity	Input B activity	Input D activity	Input E activity	Output C activity
1	0	0	0	0	0
2	0	0	0	1	0
3	0	0	1	0	0
4	0	0	1	1	0
5	0	1	0	0	1
6	0	1	0	1	1
7	0	1	1	0	0
8	0	1	1	1	0
9	1	0	0	0	1
10	1	0	0	1	0
11	1	0	1	0	1
12	1	0	1	1	0
13	1	1	0	0	1
14	1	1	0	1	1
15	1	1	1	0	1
16	1	1	1	1	0

Figure 3.7

Qualitatively consistent regulation of output C by inputs A, B, D, and E (Sansom 2008a).

inactive after B is inactive and active after B is active. Now consider the first two rows—the context of A inactivity. If C were active in the first row but inactive in the second, then B would effectively repress C. Because that is not the case, we may say that B does not effectively activate in one context but repress in another, so B influences C in a qualitatively consistent way. The same test can be applied to show that A's influence on C is qualitatively consistent by looking at rows 1 and 3, which show the context of B inactivity, and rows 2 and 4, which show the context of B activity. This test reveals that A neither effectively activates nor represses C. An input that has no effect on the function is qualitatively consistent, because the crucial test is whether it activates the output in one context but represses it in another, in which case the node is qualitatively *in*consistent. An ineffective node is not qualitatively inconsistent and therefore counts as qualitatively consistent.

Applying the qualitative-consistency test described above to the function shown in figure 3.7 reveals that it is qualitatively consistent. For example, rows 1 and 5 show us that B effectively encourages C in the context of inactivity of other nodes, and in no context does B effectively discourage C.

An attempt to produce a two-layer quantitatively consistent network capable of carrying out this function is shown in the upper right corner of figure 3.7, but, in contrast with the quantitatively consistent network shown in figure 3.6, no possible attribution of connection weight produces the function. If we assume that the network that produces this function is a two-layer quantitatively consistent network, we encounter a contradiction when trying to assign connection strengths: Row 7 shows that B does not overcome the inhibitory influence of D, and row 11 shows that A does overcome the inhibitory influence of D, so we can conclude that B's connection must weaker than A's. However, row 10 shows that A does not overcome the inhibitory influence of E, and row 6 shows that B does overcome the inhibitory influence of E, so we can conclude that B's connection must be stronger than A's. Therefore, B's connection must be both weaker and stronger than A's connection.

In order for a two-layer network to obey this function, the connection strengths would have to vary with the activity of input nodes. If the connections vary in strength depending on the activity of the other inputs, but remain either above zero or below zero, the network may still not produce a qualitatively consistent function, because of the following possible scenario: The activity of an encouraging input in one context might be accompanied by a reduction in the connection values of other encouraging inputs and/or an increase in the connection values of active inhibitory inputs, such that in one context the output is expressed when the encouraging input is not active but is not expressed when the encouraging input is active. Though the connection value for the input node in question would be positive, it would effectively inhibit the output in this context, because of the change in the other inputs' connection values.

A three-layer quantitatively consistent network that can obey this function is shown in figure 3.7. The presence of an intermediate layer of nodes allows inputs to effectively encourage the output in one context and inhibit in another, despite all connection strengths' remaining fixed. If a qualitatively consistent network may be represented by a quantitatively

	Input A activity	Input B activity	Output C activity
1	0	0	0
2	0	1	1
3	1	0	0
4	1	1	0

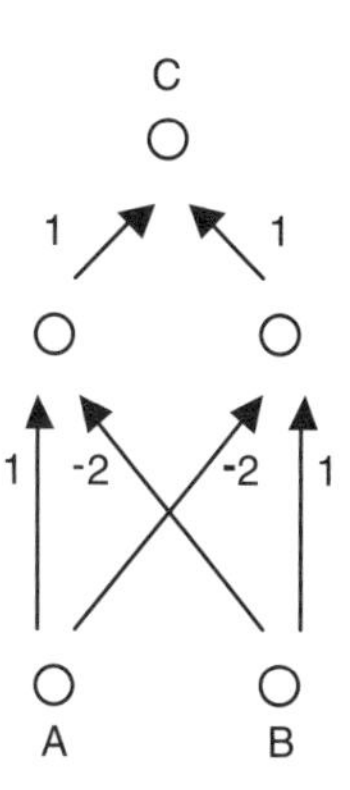

Figure 3.8
Qualitatively inconsistent regulation of output of C by inputs A and B (Sansom 2008a).

consistent network with a middle layer of units, why do we need the notion of qualitative consistency? Why not simply acknowledge that we should add an intermediate layer?

There are two reasons to keep the network two-layered but to specify that the regulation function is qualitatively consistent. First, qualitatively *in*consistent functions can also be achieved by a quantitatively consistent network with a middle layer (such as the exclusive-or function shown in figure 3.8). Nearly every transcription factor influences each of its target genes in a qualitatively consistent manner, and I will argue that this is an important source of generic order in gene regulation networks.[3] Second, if nothing limits the insertion of additional nodes in modeling gene regulation networks, then the qualitative consistency and the value of K in a model are merely matters of choice. However, Kauffman does offer such restrictions. Each node must represent a gene. It cannot represent a binding site on a regulatory region, for example. This stipulation allows my claims that gene regulation networks are qualitatively consistent and have high K to have real empirical bite. Kauffman's model and his explanation in terms of $K \approx 2$ in gene regulation networks also rely on this restriction. Such restrictions are components of the model, and to change them is to change the model. My position is that gene regulation networks should be modeled such that (i) each node represents a gene, (ii) some genes are transcription factors, and (iii) transcription factors regulate genes in a qualitatively consistent manner.

The inconsistent function (exclusive-or) shown in figure 3.8 is qualitatively inconsistent, because each node activates in one context and represses in another. For example, node B activates in the context of A inactivity (see rows 1 and 2) but represses in the context of A activity (see rows 3 and 4). In a Kauffman network, the output for each input is arbitrarily assigned, allowing all types of functions. Seventy-five percent of two-input ($K = 2$) functions are qualitatively consistent, but the proportion drops precipitously as K increases. For example, only 4.4 percent of $K = 4$ functions are qualitatively consistent.

Some transcription factors may influence the regulation of some genes in a quantitatively consistent way, either always exciting transcription to a specific degree or always inhibiting it to a specific degree, regardless of the presence of other transcription factors. Of course, the output (i.e., gene transcription rate) may vary with the presence of other transcription factors, but the influence of a quantitatively consistent transcription factor remains the same. In contrast, some transcription factors are known to act synergistically (Wray et al. 2003). The influence of some transcription factors on some genes is modified by the presence of other transcription factors. For example, some proteins have virtually no effect unless another protein is present. They may bind together before binding to the regulatory region of DNA and therefore influencing regulation (Locker 2001). Other transcription factors bind to the regulatory region and in doing so either encourage or discourage binding by other transcription factors, thereby either enhancing or inhibiting the influence of the second protein. For example, human IFN β is transcribed only if several transcription factors are present (Thanos and Maniatis 1995). In the vast majority of cases, these interactions remain qualitatively consistent.

There are some known exceptions in which an increase in the concentration of a transcription factor can take it from an activator to a repressor. For example, *Sp1* is an activator of the human folate receptor gene, but at high concentrations it interferes with the binding of another activator, *Ets*, thereby effectively repressing expression (Kelley et al. 2003). Also, *Hunchback* may go from an activator of the gene *even-skipped* at low concentrations to a repressor at high concentrations, because at high concentrations pairs of *Hunchback* molecules tend to bind to each other, thereby changing their binding capacities (Papatsenko and Levine 2008).

Cases of concentration-dependent dual regulation draw particular attention because of their exceptional nature. Do these qualitatively inconsistent transcription factors invalidate the connectionist framework for gene regulation? To the extent that these cases are exceptional, I think that the framework remains valuable, particularly with regard to general issues about order and evolvability where questions about typical features of networks are central. Later in this section, I will show how qualitatively consistent networks show much more order at high K than Kauffman networks. In the next chapter, I will show how qualitatively consistent networks are much more evolvable than Kauffman networks at high K. Were we to remove the assumption of qualitative consistency, the insight it offers into the regulation of the vast majority of genes would be lost. Scientific frameworks typically have exceptions. This is especially so in biology, owing to the great variation in phenomena. In fact, such cases of concentration-dependent regulation are inconsistent with Kauffman's model too. Rather than remain frustrated by rare exceptions, the common and (I believe) wise move is to acknowledge the exceptions and press on in the search for insight. That is what I shall do here.

Figure 3.9 shows the median of 500 first attractors of Kauffman networks of $N = 16$ with various values of K along with the median of 500 first attractors of networks with only quantitatively and qualitatively consistent node functions. (The values for quantitatively and

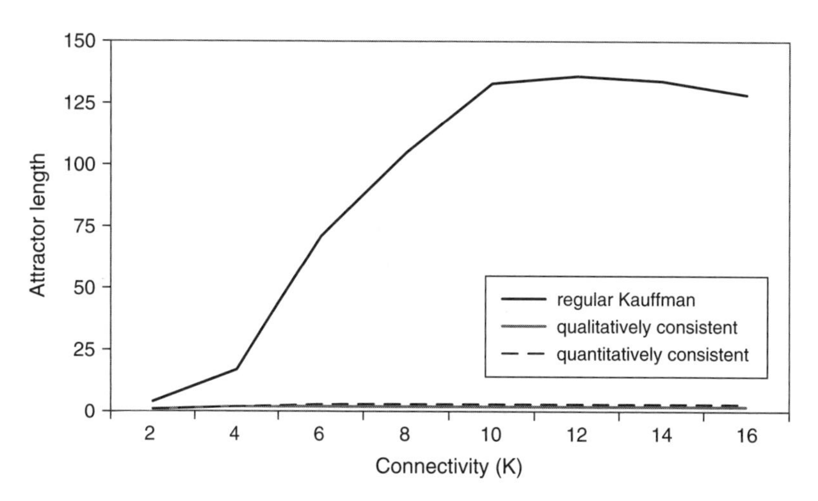

Figure 3.9
Median first attractor length; of 500 $N = 16$ networks that are $K = 2$ Kauffman networks, of 500 $N = 16$ networks that are qualitatively consistent networks, and of 500 $N = 16$ networks that are quantitatively consistent networks (Sansom 2008a).

qualitatively consistent networks are indistinguishable in this graph.) Attractor length increases with K in Kauffman networks until $K = 10$, after which it approximates a random walk around a long median. Qualitatively and quantitatively consistent networks have shorter attractors than Kauffman's networks, even at high values of K, exhibiting significantly more order.

Order is lost in high-K Kauffman networks, because the proportion of functions that are canalizing is reduced. The probability that a node in a $K = 2$ Kauffman network has a canalizing function is 0.875, but the probability is only 0.0625 if $K = 4$, and is effectively nil if $K = 6$. However, qualitatively consistent high-K networks remain orderly despite a similarly low proportion of canalizing functions. The random process of generating these qualitatively consistent networks results in all functions' canalizing in $K = 2$ networks, but the probability drops to 0.16 in $K = 4$ networks and is effectively nil for $K = 6$ networks. Quantitatively consistent networks have slightly higher probabilities for canalizing functions: 1 in $K = 2$ networks, 0.45 in $K = 4$ networks (including 0.16 with two canalizing inputs), and 0.15 in $K = 6$ networks (including 0.08 with two canalizing inputs and 0.02 with three). However, the similar order of qualitatively consistent networks clearly shows that high-K networks can be orderly despite a low proportion of canalizing functions when the functions are generated in a more biologically plausible way. (See Raeymaekers 2002.)

One factor that explains the difference in lengths of attractors of these three kinds of networks is the *average node-perturbation effect.* Consider a network with activity pattern A at t_1. It will produce an activity pattern B at t_2. Now consider activity pattern A', which is the same as A except that the activity of one node is changed (i.e., a node that is inactive in A is active in A', or vice versa). A' at t_1 would produce B' at t_2. The average node-perturbation effect is the average number of nodes that have different activities between B and B'. Networks of higher average node-perturbation effect will tend to have longer attractors. The average node-perturbation effect is determined by the average probability that the change of one input node will affect each of the nodes it influences (which I call *network sensitivity*) and by how many nodes it influences (i.e., K).

Qualitatively consistent networks are more orderly than Kauffman networks, despite not having a high proportion of canalizing functions at high K, because their functions have lower sensitivity than the randomly determined functions in Kauffman networks. Consider the first two rows of the function shown in figure 3.8. In this pair of activities, input node A is inactive. The only difference in input activity is in the activity of B. These inputs produce different outputs. Therefore, we may say that the output is sensitive to a change in the activity of B when A is inactive. The output is also sensitive to a change in the activity of B when A is active (see rows 3 and 4), to a change in the activity of A when B is inactive (see rows 1 and 3), and to a change in the activity of A when B is active (see rows 2 and 4). All told, then, the function is sensitive to all four input activity pairs of this two-input Kauffman $P = 0.5$ function, and thus is said to be 100 percent sensitive. The same analysis of the quantitative function (which must, therefore, be qualitatively consistent) in figure 3.6 shows that it is sensitive to two of the four activity pairs (50 percent sensitive). We can generate multiple quantitatively consistent and qualitatively consistent networks to induce their average sensitivity. Figure 3.10 shows that consistent functions become less sensitive with an increase in K. In contrast, it can be shown analytically that the average sensitivity for Kauffman functions is 50 percent, regardless of K. This is because the function is determined randomly, and therefore the probability that any two sets of

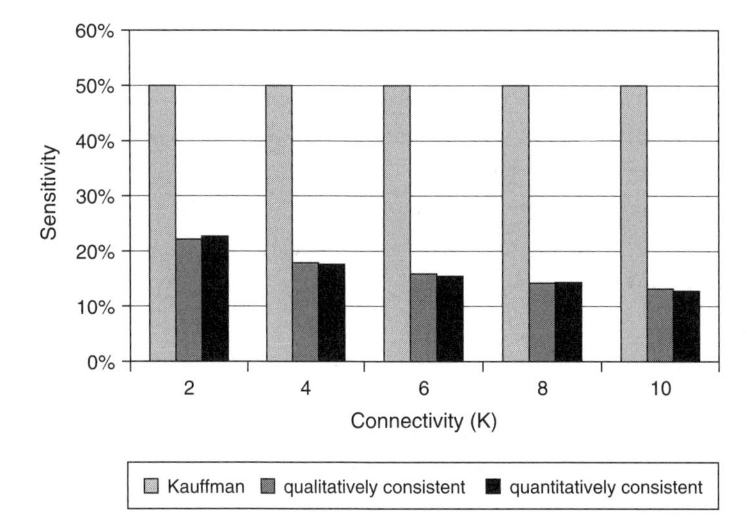

Figure 3.10

Average sensitivity; of 1,000 Kauffman networks, of 1,000 qualitatively consistent networks, and of 1,000 quantitatively consistent networks (Sansom 2008a).

inputs that vary in the activity of only one unit will produce the same output is 50 percent.

Luc Raeymaekers (2002) introduced restrictions on node functions due to inputs' being either activators or repressors (thereby producing a subset of qualitatively consistent functions) and also found that these functions reduce attractor length in networks of $K = 3$ and $K = 4$. Raeymaekers agrees with Kauffman's view that the order of development is explained by short gene regulation network attractors and finds support for this view in the shorter attractors of these networks, given that K appears to be greater than two in gene regulation networks.

The average node-perturbation effect is the likelihood that a change in input value will change the value of the output node (network sensitivity) multiplied by how many nodes it influences (K). Although sensitivity is reduced with an increase in K in qualitatively consistent networks, this is not sufficient to reduce the average node-perturbation effect as K increases. This is why Raeymaekers found that his networks of $K = 3$ had shorter attractors than his $K = 4$ networks. Raeymaekers did not investigate networks of $K > 4$. Though the attractors of qualitatively consistent networks are smaller than those of Kauffman networks of equal K, and remain short when $K = 3$ or 4, they still increase with K and N. I accept Wray's (2003) estimate that 5–15 different transcription factors for each gene is not unusual, so I do not expect the short attractors that Kauffman is looking for to explain the order of gene regulation if each node is expected to have equal numbers of activating and inhibiting input (i.e., if the networks are *balanced*). Networks of $N = 1{,}850$ and $K = 10$ might be an appropriately modified model for *Homo sapiens*, because approximately 1,850 genes are transcription factors. Figure 3.11 shows that the cycles of such qualitatively consistent balanced networks are sufficiently long that they effectively exhibit no order at all.

Raeymaekers found that networks in which the ratio of activators and repressors is not 1:1 (*imbalanced* networks) have shorter attractors. He investigated the influence of unbalancing $K = 3$ and $K = 4$ networks and found that the percentage of repressors must be below 30 or above 70 to produce significantly shorter attractors. Could gene regulation networks of $K = 10$ be orderly because of an imbalance between activators and repressors? Figure 3.11 shows that whereas qualitatively consistent $K = 4$ networks with 30 percent activators were highly orderly, those with $K = 10$

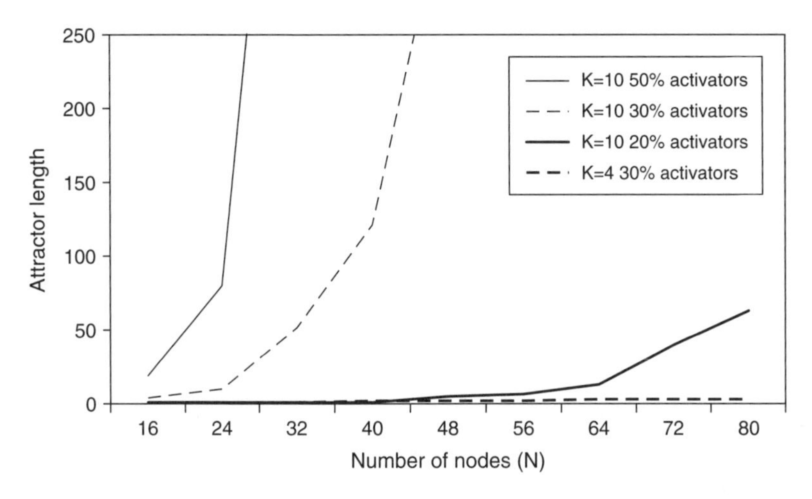

Figure 3.11
Median first attractor length of 500 K = 10 networks with 50 percent activators, of 500 K = 10 networks with 30 percent activators, of 500 K = 10 networks with 20 percent activators, and of 500 K = 4 networks with 30 percent activators (Sansom 2008a).

were not. K = 10 networks with 20 percent activators were more orderly than those with 30 percent activators, but still less orderly than K = 4 networks with 30 percent activators. Though sufficient unbalancing of a qualitatively consistent network of high K will shorten attractors to the orderly length, my results suggest that gene regulation networks must be even more imbalanced than Raeymaekers' results may suggest.

In summary, I have brought a number of biological facts to bear on Kauffman's models of gene regulation. Wray's (2003) estimate that genes are typically regulated by 5–15 transcription factors leaves low-K models of gene regulation suspect unless canalizing functions are ubiquitous or imbalance between activators and repressors is extreme. The qualitatively consistent regulation typically found in regulation by transcription factors shortens attractors in networks of high K, but need not do so by producing canalizing functions. Still, the attractors of qualitatively consistent networks of biologically plausible levels of K are not as short as those of Kauffman K = 2 networks (which Kauffman found biologically interesting) unless they are extremely imbalanced. Although many activator transcription factors have been identified in eukaryotes, which suggests that for them positive regulation by activator proteins is the predominant mode of regulation, we also know that repression is quite common (Theil et al. 2004). It is probably too soon to judge whether gene regulation networks are sufficiently imbalanced or canalized to be orderly according to Kauffman's measure, but their estimated high connectivity is not good news for Kauffman or for Raeymaekers.

3.6 What Can Short Attractors Explain?

Kauffman argues that certain facts about development are explained by the generic short attractors of gene regulation networks. Above, I used biological evidence to criticize Kauffman's explanation of why gene regulation networks should have short attractors. This leaves available the following slight variation of Kauffman's position, which could be adopted by Raeymaekers, for example, that short attractors still explain the same facts about development, but the shortness of attractors are explained by the qualitative consistency of gene regulation. In this section, I shall deny that position by criticizing the link between short attractors and those facts about development.

If we think of a cellular activity as a gene regulation network cycling through an attractor, then we can predict the number of states in a cycle of cellular activity from the number of genes in the genome (which are the nodes in a gene regulation network). If we also know how long it takes, on average, for genes to turn each other on and off (i.e., how long it takes a node to signal its neighbor in the network), then we can predict the length of time it will take for a cell to complete through an activity cycle. This Kauffman does. On his assumption that human beings have 100,000 genes, we should have attractors of 317 states.[4] Accepting that it takes about 1–10 minutes for a gene to turn on or off, Kauffman deduces that our cells should take 5–50 hours to complete an activity cycle. Kauffman takes cellular division as his example of cellular behavior. In humans, cell division takes 22–24 hours. Thus, Kauffman declares his prediction of cell cycle range to be "precisely in the plausible range of cell behavior!" (1995, p. 107).

The current view is that humans have 20,000–25,000 genes (International Human Genome Sequencing Consortium 2004). In Kauffman's $K = 2$ networks, a network with 22,500 should have approximately 150 attractors each approximately 150 states long. Therefore, attractors should take 2.5–25 hours to cycle through a cell cycle. Cell division also falls within this updated predicted range, although the fact that both predictions are consistent with the data does show some lack of precision in the model.

Kauffman also successfully estimates the length of time that cellular division will take in some other species with different size genomes (1995, p. 108). One species that he did not consider, however, is the worm *Caenorhabditis elegans,* which appears to have a relatively large genome for its complexity. *Caenorhabditis elegans* has 19,000 genes, but cell division takes only about 10 minutes during early development.[5] In contrast, Kauffman's model would expect cell division in *C. elegans* to take 138 minutes. After the first 500 minutes, cell division slows. Both the fast rate of division of an organism with a large genome and the variation of the rate of division are contrary to predictions from Kauffman's model.

On the assumption that cell types are attractors in gene regulation networks, Kauffman similarly predicts the relationship between the number of cell types of an organism and the size of its genome. For example, from the hypothesis that we humans have 100,000 genes, Kauffman predicts that we will have 317 cell types, and he measures this against the estimate that we have 256. The number of cell types is currently estimated at only 100 (Bell and Mooers 1997). On the current view that we humans have 22,500 genes, Kauffman would predict that we have 150 cell types. Again, Kauffman also successfully estimates the number of cell types in other species with different size genomes (1995, p. 109), but does not consider *C. elegans,* in which the number of cell types has been measured at 24 (Bell and Mooers 1997). Kauffman's model would expect *C. elegans* to have 138 cell types.

Kauffman also attempts to explain other developmental phenomena. Kauffman notes that during development each cell type changes into only a few different cell types. Humans have 100 cell types, but each cannot turn into any of the other 99. Instead, each can turn into only a small number of the others. The paths of differentiation are limited. In addition, in experiments we can manipulate a cell to turn into one of its neighbors in the developmental pathway, but we cannot get it to leap to another type that is far away in the developmental pathway. Finally, there are many ways to induce the same changes in cell types. Kauffman sums up these phenomena by saying that each cell is "poised" to turn into only a few cells (1993, p. 491). Kauffman explains cells' being poised with generic features of $K = 2$ network attractors on the assumption that a cell type is an attractor. If a network is cycling through an attractor, most minor perturbations to the network (such as switching one node from on to off) have no effect. The network will return to the same attractor. Some changes will result in the network falling into a different attractor, but such small changes can only move a network from any specific attractor to a small set of the other attractors. Kauffman is content that his view is supported by the fact that each cell is poised to turn into only a few other types of cell.

Kauffman admits that even if gene regulation networks are generic $K = 2$ networks, it is a large inductive step from observations about such generic networks to actual developmental phenomena. "But," he writes, "I owe you a cautionary note: whereas a great deal of work has been done on the cell cycle, we do not know enough to say more than that a powerful statistical correspondence holds between theory and observation." (1995, pp. 108–109) I discussed this issue at the end of chapter 2. At the very least, Kauffman is right to be cautious in claiming victory based on this evidence, but I think more should be said here than that caution is recommended. I wonder if Kauffman is not trading on the inductive distance from theory to phenomena to his advantage. In his prediction of cell cycle length, it seems that a gene regulation network attractor represents some cellular activity, such as cell division. But in his prediction of cell number, and in his explanation of the poised nature of cell types, it seems that a gene regulation network attractor represents a cell type, such as a liver cell. These bridging principles seem at odds. All cell types divide. Do at least the vast majority of cell types use the same cell cycle to do it? According to Kauffman's model, they must. Otherwise, the only activity that the gene regulation network could code for would be the division of each type of cell, because there are supposed to be as many attractors as cell types. If we accept that at least the majority of cell types use the same attractor for cell division, that still leaves Kauffman to wonder how a cell that is cycling through one attractor could leap to another—and not just any other, but the particular attractor that is responsible for cell division. According to Kauffman's vaunted rule that each attractor can easily access only a small number of other attractors, that attractor would appear to be unusually accessible. One response might be to suppose that each cell type has its own way of dividing included in its cell type attractor. Though this may solve the worry about cell division expressed immediately above, it implies that cells constantly divide every 5–50 hours, which does not happen for most cell types after early development.

3.7 The Order of Network Accuracy

According to Kauffman's view, cellular activity involves gene regulation networks' cycling through repeating patterns of activity. Above I have argued that the high *K* of gene regulation networks will leave gene regulation networks disorderly on Kauffman's view, unless gene regulation functions are sufficiently canalizing or imbalanced, and I have questioned the link between short attractors and various facts about development. In this section, I will discuss an important feature of development that appears inexplicable according to Kauffman's view.

According to Kauffman's framework, the only thing that a cell that remains the same type does is cycle through its attractor, which implies that there are only 150 cellular processes coded for in human beings. That number seems low in view of the complexity of complex organisms. Admittedly, the process of a cell cycle could have many sub-processes, but according to Kauffman's view such sub-processes would have to occur in a regular sequence. This would be fine if the functional demands on a cell type never varied with the environment or age of the organism, but that is not typically the case. In particular, Kauffman's view cannot explain the adaptive reactivity of cellular activity. Cells do not simply do the same thing over and over again; they seem to do the right thing in the right place at the right time. For example, *E. coli* will consume lactose only if no glucose is available. This is due to the influence of glucose on transcription factor regulation of the gene that codes for the enzyme that breaks down lactose (Reznikoff 1992).[6] In multicellular organisms, gene expression varies across life cycle, physiological condition, and environment. (See, e.g., White et al. 1999; Kayo et al. 2001; Arbeitman et al. 2002.)

Kauffman accepts that gene regulation network attractors can evolve by natural selection, but each cell type does not merely cycle through repeated activity, no matter how well designed by natural selection. Rather, cells respond adaptively to their microenvironments, doing the right thing at the right time. To effectively explain the order of development is not to explain the mere repetition of cellular activity as each cell cycles through the same attractor as the rest of its cell type, but to explain how each cell reacts adaptively to its microenvironment to allow organisms to reproduce. Put generally, not all types of order are sufficient to explain life. The world is awash with stunning examples of order that cannot reproduce, such as the patterns of activity of nonlinear systems, including snowflakes. Similarly, many human artifacts are extraordinarily ordered, but we have yet to design and build from scratch anything that can reproduce in a natural environment.

Kauffman is right that organisms are ordered systems and that gene expression is of sufficient importance to development that gene regulation networks must be ordered, too. The order of gene regulation that Kauffman focused on is the order of repeatedly cycling through an attractor. This goes hand in hand with Kauffman's view that development is fundamentally cells' repeating cycles of activity. Kauffman has also proposed that the generic order of low-connectivity autocatalytic systems of some complexity may have been crucial to the origin of life. I find the second suggestion interesting, although I will not address it in this book. It is my view that if life began as systems simply repeating activity regardless of perturbations to the environment, it has since evolved by natural selection into something importantly different that reacts adaptively to its environment.

In the next chapter, I outline what I call the connectionist framework for understanding the adaptivity and evolvability of gene regulation networks. According to the connectionist framework, each gene is directly regulated by its own little network. That network is *accurate* to the extent that the gene is transcribed only when its transcription is adaptive. According to this view, gene regulation can evolve to be accurate because of the qualitatively consistent way in which transcription factors regulate gene expression. This goes hand in hand with my view that development is not fundamentally cells' repeating cycles of activity, but rather cells' adaptively responding to changes in their microenvironments.

4 The Connectionist Framework for Gene Regulation

At any instance, each cell is faced with a microenvironment and either expresses or does not express each of the organism's genes. Given that microenvironment, each gene expression and each gene non-expression is either adaptive, maladaptive, or neutral. I consider the adaptivity of a gene regulation network to be determined by the proportion of times that it reacts adaptively to its microenvironment. A gene regulation network is *accurate* to the extent that it expresses only the right genes for each microenvironment. The problem of the evolvability of gene regulation networks becomes the question of how gene regulation networks can evolve to allow such accuracy. I propose that it is the similarity of these networks to connectionist networks that explains their evolvability.

Whether a gene's expression in a microenvironment is adaptive or not is a complex biological fact, determined by the environment and by other facts of development. A further complicated is that it may be adaptive for the organism but not for the group of which that organism is a member. There is a fact of the matter, as complex and as difficult to discover as it may be. Discussion of adaptive gene expression is supposed to boil down all the developmental and environmental complexity to the gene level. This must be done if we are to talk about the evolution of genes, but it necessarily overlooks important details of why the facts are the way they are. This is the price of taking a gene-level perspective that was discussed at the end of chapter 1 and will be defended in chapter 5.

4.1 Adaptive Gene Regulation

An accurate gene regulation network responds adaptively to the microenvironment of each cell. In multi-cellular organisms, the microenvironment

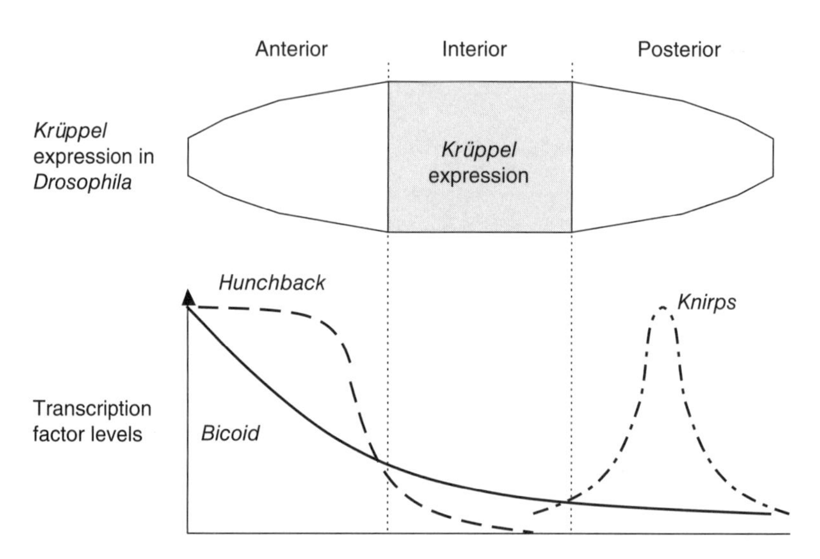

Figure 4.1
Regulation of *Krüppel* in early development of *Drosophila melanogaster* (Sansom 2008b).

of a particular cell can include positional information (Wolpert 1970). For instance, in early stages of the development of *Drosophila melanogaster*, the gene *Krüppel* is expressed only in the interior segment. This is due to its regulation by three transcription factors: *Hunchback*, *Knirps*, and *Bicoid*. Figure 4.1 shows the levels of these transcription factor proteins in the anterior-segment, interior-segment, and posterior-segment microenvironments at this stage of development. These transcription factors interact in this network in such a way that the levels of *Hunchback* and *Knirps*, relative to the level of *Bicoid*, determine the thresholds between regions of expression and non-expression of *Krüppel*. The regulation of *Krüppel* shown in figure 4.1 is adaptive, because it is crucial to normal development of *Drosophila melanogaster* that *Krüppel* be expressed only in the interior segment. *Krüppel* itself produces another transcription factor that plays a role in regulating the expression of other genes, and its expression in only the interior segment contributes positional information too.

How did the regulation of *Krüppel* evolve to be so highly adaptive? Mutations to the regulatory region of a gene can add, remove, or in other ways alter transcription factor binding sites. In doing so, they evolve the regulatory relationships between specific transcription factors and genes, which is equivalent to saying that the gene regulation network can evolve across generations. Several studies suggest that much evolutionary change occurs in gene regulation as transcription factors gain and lose influence (Doebley and Lukens 1998; Carroll 2000; Stern 2000; Tautz 2000; Thiessen et al. 2000; Purugganan 2000; Wray and Lowe 2000; Carroll et al. 2001; Davidson 2001; Wilkins 2002). In my search for a general framework for understanding the adaptivity and evolvability of gene regulation networks, I turned to an old tool of artificial intelligence: the connectionist network. In the next two sections, I will discuss the many similarities and the big difference between connectionist networks and gene regulation networks. In subsequent sections, I will argue that this difference has little significance when it comes to drawing certain lessons about the adaptivity and evolvability of gene regulation networks that I offer from the connectionist framework.

4.2 Similarities between Connectionist Networks and Gene Regulation Networks

Connectionist networks (also known as parallel distributed processors and neural networks) can be described in terms of the following eight aspects (taken from page 46 of Rumelhart et al. 1986):

- a set of processing units
- a state of activation
- an output function
- a pattern of connectivity among units
- a propagation rule for propagating patterns of activities through the network of connectivities
- an activation rule for combining the inputs impinging on a unit with the current state of that unit to produce a new level of activation for the unit
- a learning rule whereby patterns of connectivity are modified by experience
- an environment within which the system must operate.

Consider the simple network illustrated in figure 4.2. Each circle is a unit in the network. Each unit has an activation level at each time, $a_u(t)$.

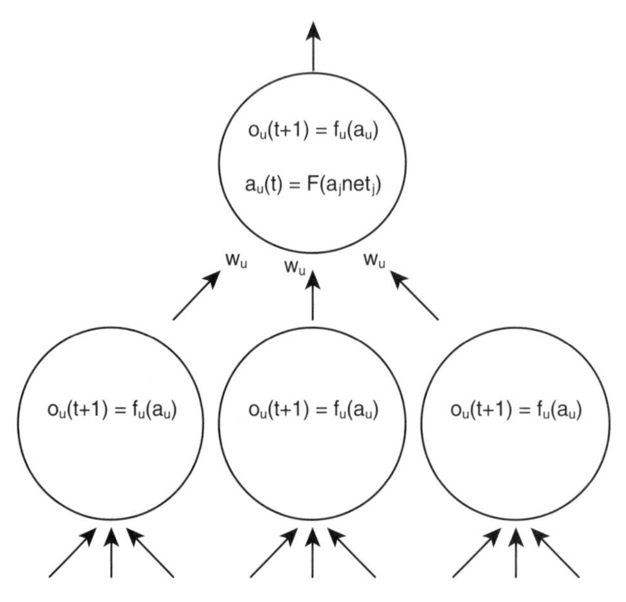

Figure 4.2
Generic description of a connectionist network (slightly modified from Rumelhart et al. 1986, p. 47) (Sansom 2008b).

The output of each unit at the next period of time, $o_u(t + 1)$, is a function f of its unit activation level. The output passes along unidirectional connections indicated by arrows. The weight of each connection is a real number, w_u, which determines the magnitude of the input of the one node from the output of the other. All inputs to a node are then combined by an operator (usually addition, $\Sigma o_u w_u$ = net$_j$) to determine the activity of that node by function F: $a_u(t) = F(a_j \text{ net}_j)$.

Engineers have put connectionist networks to a great variety of tasks. For example, Sejnowski and Rosenberg's (1987) NETtalk was a network that was capable of pronouncing English text. The process typically begins with the engineer determining the general architecture of the network. This requires determining the number of input units and the aspect of the environment they represent, the number of output units and what they represent, and how many hidden nodes there are between the inputs and outputs. For example, NETtalk's input units were in seven groups, each representing a letter in a seven-letter window of text, and its output was code for a speech synthesizer.

A network's level of performance is determined by how often it gives the correct output for the input it receives for each member of a training set, which depends on the network's architecture and on the weights of the connections between the nodes. A network with randomly assigned connections will achieve generic performance, but its performance can be improved by changing the weights of the connections. At first NETtalk produced random noise, but with "training" it progressed to babble, then to nonsense collections of sounds used in English language, then to understandable English.

As connections are changed to improve performance, intermediate nodes gain the capacity to represent. For example, NETtalk managed to distinguish vowels from consonants, then to distinguish between particular vowels and between particular consonants. Acquired representations of internal nodes tend to be distributed across patterns of the activity of multiple nodes, rather than the activity of an individual node representing something by itself.

Networks are trained on the basis of their response to a set of examples of the task they are to perform. Phonetic transcripts from the speech of a child and *Merriam-Webster's Pocket Dictionary* were used to train NETtalk. Engineers typically use supervised learning to modify their networks, which require immediate and detailed feedback to determine which connections are contributing to the success of the network for each example (known as solving the "credit-assignment" problem) and increasing their relative strength. These changes are *directed*, because they make changes that are likely to improve the network's performance. NETtalk was trained in such a directed way. This method does not require knowing how to get from the input to the right output. In a sense, the network figures that out for itself as it is trained.

Weights and/or architecture in connectionist networks can also be effectively modified using evolutionary algorithms—i.e., artificial selection processes (Holland 1975; Koza 1992; Schwefel 1995; Schlessinger et al. 2005). Typically this is done by repeatedly testing the performance of a population of networks using the training set and producing the next generation by making random mutations to the most successful networks. The changes made to individual networks in this process are not directed, but the population of networks can be expected to improve performance above generic levels because direction comes in the fact that the most successful networks are selected at each generation. This method requires less knowledge than other methods—only the level of performance over the entire evaluation period. For this reason, Nolfi and Parisi (2002) claim that it tends to produce better results when the detailed feedback needed to solve the credit-assignment problem isn't available.

Schlessinger, Bentley, and Lotto (2005) investigate the evolution of architecture and connections in networks in Mosaic World, a two-dimensional artificial life environment. Input units represent aspects of each "critters" environment, and output units represent critter behavior, such as mating, eating, and moving. The representational content of these networks are acquired over the evolutionary process. For example, input units are modified to enable the evolution of vision. Understanding the exact nature of acquired representations is typically not straightforward. In examples like this from artificial life, the representational contents of aspects of the network are determined by the functional role that they play in contributing to the adaptivity of the critters.

Accepting networks produced by artificial selection as satisfying Rumelhart, Hinton and McClelland's criteria for being connectionist networks requires accepting that artificial selection counts as a learning rule, whereby patterns of connectivity are modified by experience. While there are significant differences between supervised learning and artificial selection, both methods are used to achieve a connectionist network that produces better-than-generic performance. I will refer to networks that are artificially selected for performance as connectionist networks.

Figure 4.3 is an example of the above type of network shown in figure 4.2, with certain variables determined. It describes the regulation of the gene *Krüppel* in early development of *Drosophila melanogaster*. The activity of input units *H*, *B*, and *k* are the levels of the transcription factors *Hunchback*, *Bicoid*, and *Knirps*, respectively and the output, *K*, is *Krüppel* expression. The activity of each input is equal to some aspect of the network environment, which is measured on a scale between 0 and 1. *Hunchback* and *Bicoid* appear to represent proximity to the anterior segment and *Knirps* proximity to the posterior segment. The output of all input nodes is equal to its activity at the previous time. The weight of the left and right connections is -1 and the weight of the middle connection is 1. The inputs to the output node are summed to determine its activity. The network output is either 1, if the in sum of the input node's outputs is

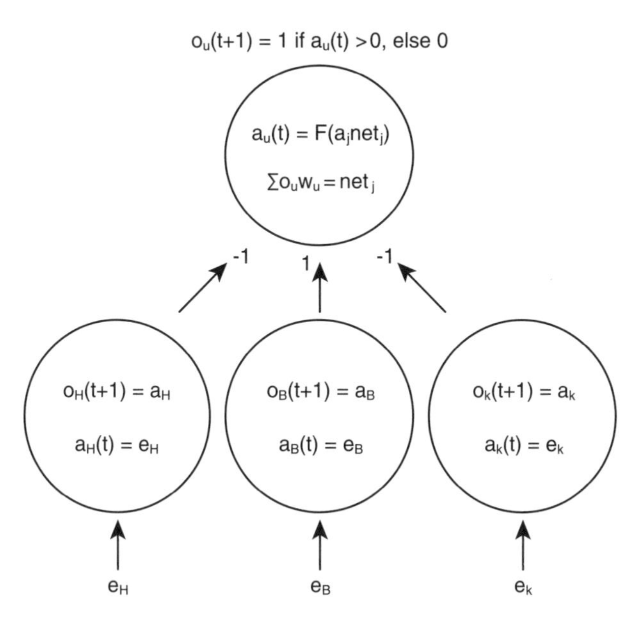

Figure 4.3
Connectionist regulation of *Krüppel* (Sansom 2008b).

greater than zero, or the network output is zero. Figure 4.4 shows the activity of a network in the anterior, interior, and posterior segments in early development of *Drosophila melanogaster.*

The ability to show a gene regulation network as an instantiation of Rumelhart, Hinton and McClelland's generic description of connectionist networks indicates the strikingly similarity between gene regulation networks and connectionist networks designed by artificial selection.

I propose the connectionist view of gene regulation networks, which is primarily proposed to explain the adaptive reactivity of gene expression of a cell to its microenvironment (i.e., *network accuracy*). Different transcription factors are present in different cellular microenvironments within an organism. They can either activate or repress transcription rates of specific genes by binding to their regulatory regions. The variations in transcription factor presence across microenvironments, along with the way they influence the transcription of each gene, determine in which microenvironments that gene is transcribed, which is the gene's *expression profile.* Mutations to the regulatory region of a gene may result in a transcription factor no longer influencing that gene or another transcription factor

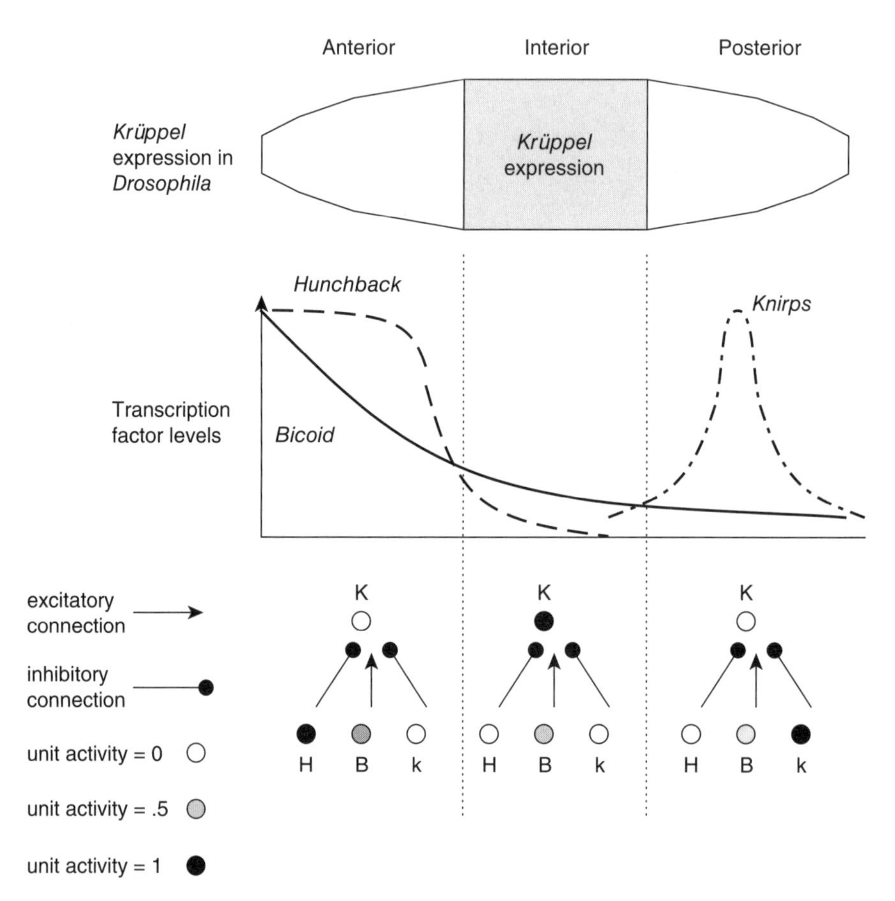

Figure 4.4
Krüppel regulation network activity in early development of *Drosophila melanogaster* (Sansom 2008b).

gaining influence. The mutation will be adaptive if the gene's new expression profile is more adaptive than its expression profile before the mutation.

Kauffman's order of attractors deals explicitly with the behavior of the entire network (as simplified in figure 3.2). At the beginning of section 3.5, I suggested that the attractors of the entire gene regulation network will be fully determined by the attractors of the smaller, but still quite large, transcription factor regulation network. My view concentrates on the small network of transcription factors that directly regulates a particular *structural gene* (i.e., any gene that does not produce a transcription factor).

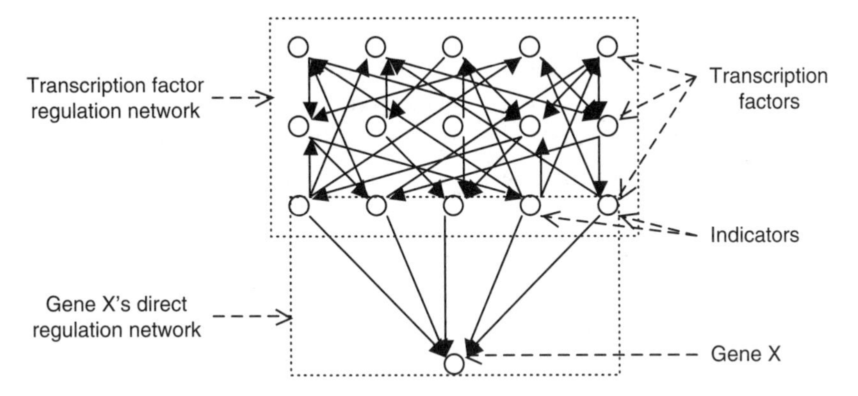

Figure 4.5
The connectionist view of gene regulation networks (Sansom 2008a).

However, its accuracy is about the transcription factor regulation network as well.

Figure 4.5 represents the connectionist view of the regulation for gene X. The transcription factor regulation network determines the presence or absence of each transcription factor in each microenvironment. If a transcription factor is present in most of the microenvironments in which X expression is adaptive, and absent in those in which it X non-expression is adaptive, then that gene is a relatively accurate potential indicator of X expression. If that transcription factor is an activator for that gene, then that potential is realized and it is an accurate X expression indicator.

If all gene X's activators are accurate expression indicators, and all its repressors are accurate non-expression indicators, then we would expect that the direct gene regulation network is accurate, because it will result in that gene having an adaptive expression profile. To say that gene X's direct regulation network is accurate is not to make a claim about only that gene and how its transcription factors directly influence that gene's transcription, because the indicator accuracy of those transcription factors is due in part to other transcription factors that directly regulate them and yet others that directly regulate those transcription factor etc. (i.e., the whole transcription factor regulation network), because they determine in which microenvironments the directly regulating transcription factors appear. Network accuracy is also contingent on the effect of the structural gene because this determines the adaptivity of its expression in each microenvironment. The accuracy of the regulation networks of each structural gene together determine the accuracy of the entire gene regulation network. The notion of accuracy can be applied to the small network that directly regulates any transcription factor too, but this would be a derivative of the gene regulation of the structural genes that are regulated by that transcription factor, because they determine in which environments the expression of that transcription factor is adaptive.

4.3 The Difference between Connectionist and Gene Regulation Networks

In connectionist networks, the function between the set of input nodes and the nodes that they directly influence is "linearly separable." I call such networks *quantitatively consistent* network to distinguish them from *qualitatively consistent* networks. A network is quantitatively consistent if its connections maintain the same value, regardless of the activity of other inputs. A network is qualitatively consistent if no input unit directly activates an output in one context and represses that same output in another; otherwise it is *qualitatively inconsistent* (see section 3.5).

The consistency or inconsistency of a network suggests something about the causal mechanism involved. Consider two ways for a department to choose a new colleague. One system is to have each member of the department interview the candidate and vote, without discussing the matter with any other member. This may be represented as a quantitatively consistent connectionist system, because each department member acts independently. The second way is to have departmental meetings, where the candidate is discussed and members can influence each other, perhaps by saying who they are going to vote for and why. I do not wish to begin to evaluate which is the best system for selecting new members of a department. My point is that not every system has consistent connections and therefore should be modeled as a connectionist system.

I propose connectionism as a framework for gene regulation. It is more than the claim that gene regulation networks can be simulated using connectionist networks. Just about anything might be so simulated with some degree of success, from the spread of fire to an electoral process, despite connectionism telling us nothing about the mechanisms of the spread of fire. A connectionist framework has been applied to help understand the brain (e.g., MacDonald and MacDonald 1995, O'Brien and Opie 1999) and I am proposing that it be applied to gene regulation too.

I will show that because qualitatively consistent gene regulation networks are sufficiently similar to quantitatively consistent connectionist networks that some general lessons about gene regulation network adaptivity, evolvability, and evolutionary trends can be learned by examining the generic features of connectionist networks. Some insights follow from the connectionist literature, and some explanations of my own were produced using the framework.

The connectionist framework for gene regulation comes with qualifications. Rare exceptions to gene regulation being qualitatively consistent were discussed in section 3.5. Additionally, there are some aspects of gene expression, such as methylation of DNA and alternative splicing of RNA that have yet to be incorporated. DNA methylation prevents transcription of specific regulatory regions. This can be seen as removing connections from the network. Somatic mutation and alternate splicing in immune system cells may result in a slightly different gene regulation network for some cells in an organism. Both somatic mutations and methylation patterns are passed on to daughter cells during development. Does this mean that different lineages of cells have different regulation networks? This is an interesting issue and a complete theory would certainly take this cell lineage effect into account. However, I will continue to say that all cells of an organism share the same regulation network for modeling simplicity. I think that this is a justifiable simplification because differently methylated networks in an organism may still have many connections in common (i.e., methylation can remove connections but not create new ones) and most regulation of gene expression occurs at the rate of transcription initiation (Latchman 1998; Carey and Smale 2000; Lemon and Tjian 2000; White 2001; Wray et al. 2002). Alternative splicing of RNA also complicates matters because it allows the same transcribed region to be responsible for the production of different proteins. Here too, the protein produced is tightly constrained by the region transcribed.

How much insight does the connectionist framework offer? I use the framework to offer some analyses of gene regulation phenomena below, but another way to measure my case for adopting the connectionist framework is to consider its advantages over the most developed alternative, which is Kauffman's (1993, 1995) theory discussed in chapter 3.

I consider the debate about whether to use the connectionist framework or Kauffman's to be analogous to the debate about applying the connectionist or the classical frameworks to the brain (e.g., MacDonald and Macdonald 1995). Accordingly, this is best judged after both perspectives have had time to develop and prove how fruitful they can be.

4.4 The Adaptivity of Gene Regulation

The expression profile of a gene describes in which microenvironments it is expressed during development. For any gene, there is an optimally adaptive gene expression profile. The regulatory network for that gene is adaptive to the extent that it matches that profile. The more thresholds between ranges of expression and non-expression across microenvironment space in the optimal expression profile, the greater its complexity, and the more difficult it is to match. It is easy to design an adaptive network for a gene that should always be expressed. All you need is a network with numerous various activating transcription factors and few (if any) repressing connections. This may describe the regulation of "housekeeping genes" which are expressed unless there are specific conditions, such as heat shock and starvation (Pirkkala et al. 2001). In contrast, a gene with a complex adaptive expression profile is more difficult to design.

Any gene's optimal expression profile may be exactly the same as a transcription factor's expression profile. In such a case, that transcription factor is all that is needed to regulate the gene in question. However, the more complex the adaptive gene expression pattern, the less likely this is. Most genes are not transcription factors and different genes have different optimal expression profiles. Therefore, the optimally adaptive gene expression profile for most genes, at the very least, will not be matched by the expression profile of any single transcription factor. Instead, natural selection finds combinations of transcription factors whose presence and absence is closer to the optimal expression profile of each target gene and the average transcription factor is involved in the regulation of tens to hundreds of genes (Wray et al. 2003). For example, in *Drosophila melanogaster*, Ubx isoform Ia directly regulates an estimated 85–170 genes (Mastick et al. 1995).

Combinations of transcription factors can adaptively regulate target gene expression, because the effects of transcription factors that regulate a gene are integrated in a qualitatively consistent way. Figure 4.4 (above) represents how transcription factors together determine one region of expression of *Krüppel* expression in early development of *Drosophila melanogaster*. Regulation of genes expressed early in development typically integrates transcription factors that specify temporal and spatial regions (Davidson 2001; Wilkins 2002). Other genes are regulated such that they are only expressed in response to specific hormonal, physiological, or environmental cues (Shore and Sharrocks 2001; Benecke et al. 2001).

Consider the following two classes of connectionist gene regulation networks. All units are Boolean (i.e., active or inactive). The activity of the input units determines whether the target gene X is expressed (call this output X) or not (call this output not-X). These two possible outputs are exclusive and exhaustive. The gene regulation network is adaptive to the extent that it allows the expression of X in microenvironments where the production of X is adaptive (X microenvironments) and it disallows expression where the production of X is maladaptive (not-X microenvironments). The default for these networks is to not express the target gene.

All microenvironments are either X microenvironments or not-X microenvironments. Some transcription factors will be present in a greater proportion of X microenvironments than not-X microenvironments. These transcription factors are X indicators. The accuracy of an X indicator is a measure of the proportion of its presence in X microenvironments and absence in not-X microenvironments. It is not appropriate to model any X indicator (or not-X indicator) as 100 percent accurate, because of the small number of transcription factors and variety in target gene optimal expression profiles discussed above. Therefore, assume that all X indicators and not-X indicators are 75 percent accurate (e.g., X indicators are present in 75 percent of X microenvironments and absent in 75 percent of not-X microenvironments). The errors of one input unit are independent of those of other units. This means that an X microenvironment, in which one X indicator is erroneously absent, is no more or less likely than any other X microenvironment to have other X indicators erroneously absent as well. X indicators activate the output and not-X indicators repress it.

In the quantitatively consistent connectionist class of networks, the strength of the connections from X inputs varies between 1 and 0 (not-X inputs between –1 and 0), but each remains the same across the activity of other inputs. In the qualitatively consistent class, connection strengths vary based on the activity of the other input units. However, because they are qualitatively consistent, the values can never change such that any node activates in one context and represses in another. As discussed in section 3.5, this requires that network connections do not change such that the activation of an excitatory input is accompanied by a weakening of other excitatory connections or strengthening of inhibitory connections so that the output is turned off (*mutatis mutandis* for inhibitory inputs).

The accuracy of a network is the average of the proportion of X microenvironments in which the network will output X and the proportion of not-X microenvironments in which the network will output not-X. Figures 4.6 and 4.7 show the accuracy of samples of these types of networks. One hundred networks of each variety listed along the horizontal axis were randomly generated. The output of the network for every possible combination of inputs was determined, along with the probability of each combination occurring in an X microenvironment and not-X microenvironment. On the assumption that all indicators are 75 percent accurate, the likelihood of most combinations of inputs changes if the situation is an X microenvironment as opposed to a not-X microenvironment. For example,

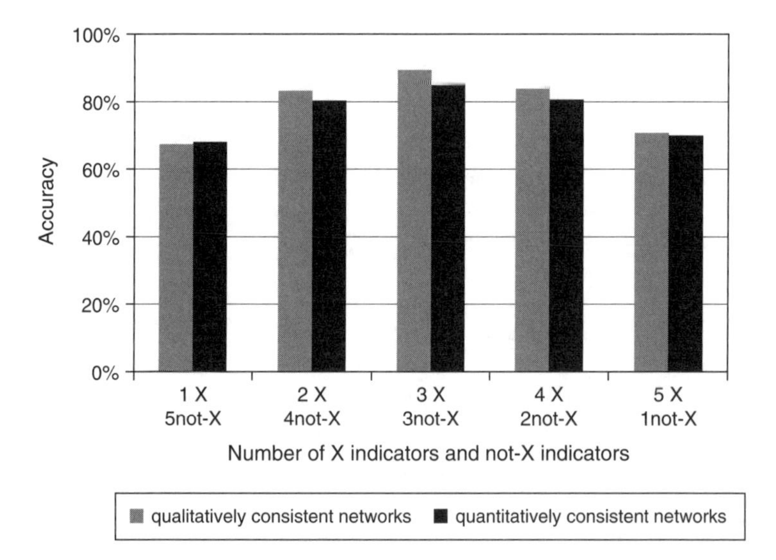

Figure 4.6
The adaptivity of balanced connectionist networks (Sansom 2008a).

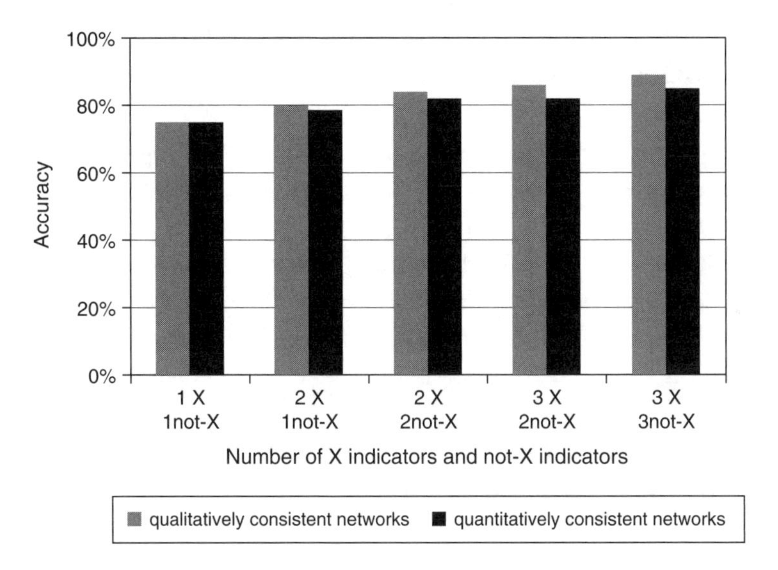

Figure 4.7
Adaptivity of high *K* balanced qualitatively consistent and quantitatively consistent networks (Sansom 2008a).

the presence of only X indicators and absence of all not-X indicators is more probable in an X microenvironment than a not-X microenvironment. The network's accuracy across all X microenvironments were determined by finding all input combinations that result in an X output and summing the probabilities of each in an X microenvironment (mutatis mutandis for network accuracy in not-X situations). The overall network accuracy is an average of the two.

A network is balanced if it has an equal number of excitatory X indicators and inhibitory not-X indicators. Balanced qualitatively and quantitatively consistent networks tend to be more accurate overall (figure 4.6). Therefore, all things being equal, it is most adaptive to have a balanced network. All things may not be equal, though. For example, if Xing in not-X microenvironments is not very harmful, but not-Xing in X microenvironments is, then a network imbalanced toward more X indicators is more adaptive, because it is more likely to X than not-X, so while less accurate overall, it is still more accurate in X microenvironments. Additionally, X microenvironments may be more common than not-X microenvironments, putting a greater premium on being accurate in X microenvironments than not-X microenvironments. With these possibilities acknowledged, this discussion assumes that both types of inaccuracy are equally costly.

Accuracy increases in qualitatively and quantitatively balanced connectionist networks as the number of input nodes increases (figure 4.7). The fundamental reason for the increased accuracy was provided by Condorcet's jury theorem in 1785 (McLean and Hewitt 1994, pp. 131–138), which is much discussed in political theory. If a group of people are voting together on a decision, if each person has a higher than 50 percent probability of voting for the best outcome, the chance of the majority voting for the best outcome approaches 100 percent as the number of voters increases. This is because in each situation the accurate majority will cover for the inaccurate minority. The same is true of balanced connectionist networks.[1] Another way to think about this is that in certain situations the output of the network is overdetermined. An output is overdetermined if it is not contingent on the presence of any single indicator, so the error of any single indicator makes no difference.

These results rely on the assumption that indicators are independent. The adaptivity of redundant parts is always dependent on the failure of each part's being independent of the failure of others. Complete independence of indicators is unlikely in biological systems, where parts affect other parts. Nevertheless, I expect that there is some indicator independence. A great many things go on over the course of development of a complex organism. Much of this activity involves transcription factors that could change gene expression profiles of genes with the right mutation of their regulatory regions. As these regions mutate, natural selection will have many opportunities to evolve arbitrary links between formerly unrelated processes and sub-processes, because they are all controlled by the common currency of transcription factors. Each transcription factor is such that its role will be quite specific, but that role can change with mutation, linking it to new processes and removing it from significance to others. Such mutations will tend to result in diversity of transcription factor expression profiles. Thus, I expect that indicators can be relatively independent. Not only can indicators be independent, natural selection will favor independent indicators. If natural selection is presented with two similarly accurate additional X indicators, but one makes the same sorts of mistakes as the other X indicators already in the network while the other is independent, then natural selection will favor the independent indicator, because that will produce a more accurate network.

In section 3.5, I discussed how improving network balance reduced generic network order according to Kauffman's measure (i.e., attractor length). Balanced networks are more sensitive and have longer attractors than randomly determined networks, which have many unbalanced nodes. In contrast, balanced networks are more orderly according to my measure of order (i.e., network accuracy). I also discussed how gene regulation networks have sufficiently high values of *K* to take them out of the ordered realm by Kauffman's measure of order. In contrast, figure 4.7 shows that increasing *K* (i.e., increasing the number of inputs while maintaining balance) improves network accuracy.

I have shown the adaptivity of a high number of inputs in gene regulation networks. While the same logic can be applied to engineered networks, engineers face a risk when using too many inputs when training their networks, using either supervised learning or artificial selection. Both training methods use a set of examples. Networks are modified to do a good job on those examples with the expectation that they will translate that success to similar cases when the network is put to real use. The risk is that the network will become "overfitted" to the training set, so that it is excellent at producing the correct output for those examples but not very good at doing so for real world examples that differ even slightly. This risk is increased as the number of inputs goes up.

The risk of overfitting is reduced to almost nothing in a network evolved by natural selection, because the "training set" of the situations faced by naturally selected ancestors is enormously greater than the number of situations that will be faced by the network of an individual. It is true that if an organism faces a substantially novel environment, or has another mutation that changes the microenvironments of its cells, the gene regulation network will be presented with microenvironments to which it has not evolved by natural selection. That is not a case of overfitting, but rather a case of a network evolving by selection for one type of task and then presented with another. No training system does well in such a case, including natural selection, because it never anticipates the future, but only selects on past performance.

I have already discussed how nodes can gain representational content in connectionist networks that we engineer using supervised learning or artificial selection. We may be interested in determining the representational content of transcription factors. For example, if one looks at the expression pattern of the transcription factor *Krüppel*, shown in figure 4.1, it is tempting to say that it represents the interior segment of this early stage of development of *Drosophila melanogaster*. Lessons from biology should give us pause, though, because transcription factors can be expressed in different microenvironments in the same species and might represent more than one content. Lessons from connectionism suggest that representation of transcription factors may be more complicated still. In connectionist networks, representation can be realized in terms of the activity of multiple units. The units may only represent anything that might increase fitness together, but nothing individually. This suggests that we investigate the following sorts of phenomena. What other transcription factors co-regulate target genes that are regulated by *Krüppel*? Does each maintain its excitatory and inhibitory status across this set of target genes? If so, what is the boundary of gene expression of all these target genes?

4.5 The Evolvability of Gene Regulation

The probability of a mutation to a natural system being adaptive is nearly always going to be lower than 50 percent. However, some types of mutation may still have a better chance than others and we can describe such mutations as highly evolvable. While measuring the evolvability of gene regulation networks remains a complex theoretical and empirical challenge, empirical evidence suggests that they have evolved a great deal over the course of evolution. Regulatory regions may have evolved more than coding regions (Gerhart and Kirschner 1997; Stern 2000; Carroll et al. 2001; Wilkins 2002). For example, a survey of twenty regulatory regions in mammals showed that probably only two-thirds of binding sites in humans are functional in rodents (Dermitzakis and Clark 2002). In addition, different gene expression has been linked to various phenotype differences, reviewed in Wray et al. (2003), including (1) anatomy (Burke et al. 1995; Averof and Patel 1997; Stern 1998; Wang et al. 1999; Lettice et al. 2002); (2) physiology (Abraham and Doane 1978; Matsuo and Yamazaki 1984; Dudareva et al. 1996; Sinha and Kellog 1996; Stochaus et al. 1997; Segal et al. 1999; Lerman et al. 2003); (3) behavior (Trefilov et al. 2000; Saito et al. 2002; Caspi et al. 2002; Enard et al. 2002b; Fang et al. 2002; Hairi et al.

2002; Meyer et al. 2002); (4) polyphenism (Brakefield et al. 1996; Abouheif and Wray 2002); and (5) life history (Allendorf 1982, 1983; Anisimov et al. 2001; Streelman and Kocher 2002). This is evidence that gene regulation has non-neutrally evolved a great deal over the course of evolution. This alone is somewhat suggestive of gene regulation networks' being highly evolvable. It is necessary to understand how gene regulation networks are organized to show how they can evolve (Davidson 2001; Wagner 2001).

Schlessinger et al. (2005) investigate the evolvability of structural mutations to connectionist networks. Of the various architectural mutations that they consider, the one most likely to improve network function was the addition of a node with only a few randomly generated connections. Translating this result from connectionist literature to gene regulation networks suggests that mutations that add a transcription factor to the regulation of a gene are more likely than other types of mutations to be adaptive. Why are such mutations are highly evolvable?

Consider the question, how accurate must an X indicator be in order for its addition to improve the accuracy of a network? All things being equal, the answer depends on whether the addition improves the balance of the network or not (i.e., brings the number of X indicators closer to the number of not-X indicators). Broadly speaking, if the indicator worsens network balance (e.g., if it already had more X indicators than not-X indicators), then it must be more accurate than the other indicators in the network to improve network accuracy. But if the indicator improves balance, it need be only more than 50 percent accurate to improve accuracy. These results indicate that connectionist networks are highly evolvable, but do not yet explain why this is so. In this section, I shall use the connectionist framework to offer a preliminary analysis that shows why gene regulation networks meet certain general requirements for effective design by natural selection.

Charles Darwin famously believed that traits evolve gradually by natural selection. Mutation is a generic feature of reproduction; any natural reproduction system will have a mutation rate greater than zero. Though random changes to well-designed systems tend to be harmful, some fraction of random mutations are adaptive, and the smaller the effect (or the fewer the effects) of a mutation, the more likely this is. It is always unlikely that a mutation will be adaptive, but a smaller mutation still is more likely to be adaptive than a larger mutation. (See chapter 1.)

The success of artificial selection in modifying connectionist networks to improve performance suggests that connection strengths are evolvable. Generally, the more gradually a well-designed system mutates, the more evolvable it is. Below, I will use the connectionist framework of gene regulation to show how gene regulation networks can evolve gradually and how this evolvability may be increased as the number of input nodes is increased.

In section 3.5, I described the notion of *network sensitivity* (that is, the average probability that changing the activity of one node will affect each of the nodes it influences), and I used that concept to explain why qualitatively and quantitatively consistent networks are less sensitive to perturbations than Kauffman's networks and why they become less sensitive with an increase in the number of inputs. This concept was introduced within the context of Kauffman's framework, and it explained why qualitatively and quantitatively consistent high-*K* networks have relatively short attractors. Within the context of the connectionist framework, the low sensitivity of these networks explains their evolvability in terms of network accuracy.

When assessing whether gene regulation networks are evolvable, one must measure the change to the network after mutation. In order to do this, one must make assumptions about which differences make a difference. I assume that gene regulation networks are systems that have the function of delivering adaptive gene-product outputs for each microenvironment faced by cells in the organism. Therefore, the size of an average mutation to a network should be measured by the proportion of possible input activity states that will have their output changed after the mutation. The smaller the proportion, the more gradual mutation will tend to be and, therefore, the more evolvable the system will be.

A microenvironment is *sensitive* to a particular mutation to a gene regulation network only if that mutation will change the expression of that gene in that microenvironment. For example, if we consider the three microenvironments depicted in figure 4.4, we see that each is sensitive to only one kind of mutation involving the loss of a regulatory connection. The anterior microenvironment is sensitive only to a loss of the inhibitory connection from *Hunchback*, the interior microenvironment is sensitive only to the loss of the excitatory connection from *Bicoid*, and the posterior microenvironment is sensitive only to the loss of the inhibitory connection from *Knirps*. Thus, any of these mutations would change the expression of *Krüppel* only in one of these microenvironments.

As I discussed in chapter 3, Kauffman's networks of any size have 50 percent sensitivity, but qualitatively and quantitatively consistent networks become less sensitive as they increase in the number of inputs. (See figure 3.10.) Consider a mutation that adds an activator. In order to be adaptive and improve the network's accuracy, it need only be highly accurate within microenvironments that are sensitive to the mutation, because in only these microenvironments will the mutation make any difference. If we assume that different transcription factors that might be added to a network vary in the microenvironments in which they tend to be present, the smaller the sensitive range of a mutation, the more likely that the added transcription factor will be highly accurate within that range. As the sensitive range approaches zero, the probability that an added transcription factor is accurate within that range approaches 50 percent.

The quantitatively consistent connectionist networks investigated in chapter 3 remain slightly less sensitive than qualitatively consistent networks. In those qualitatively connectionist networks, the strength of each connection is contingent on the presence of all the other inputs. This is not the case in gene regulation networks, in which the strength of a connection from one input will either be independent of all other inputs or dependent on only a small number of them. For example, consider a mutation that adds an inhibitory transcription factor that effectively prevents an excitatory transcription factor from binding. Such a mutation may have an effect only in microenvironments in which the excitatory transcription factor is present, so the sensitive range for such a mutation is smaller than for a mutation that would add an inhibitory transcription factor whose influence was independent of any other transcription factors. As a result, gene regulation networks should be more evolvable than generic qualitatively consistent networks.

This connectionist theoretical analysis of gene regulation networks shows that mutations to gene regulation networks that involve many transcription factors will have limited effects, making the networks highly evolvable. This stands in stark contrast to Kauffman's expectation that networks of higher K will be less evolvable than networks of lower K.

4.6 Evolvability versus Robustness

Robert Leclerc (2008) has presented a number of simulations that suggest that evolution favors low-K networks of an order similar to Kauffman's favored $K = 2$. Whereas Kauffman's work is based on cycles of repeated activity, Leclerc's work is based on network accuracy. Leclerc finds a selective advantage in low-K networks because of their greater robustness. A robust network is more resilient in the face of perturbation or mutation than a less robust network—that is, mutation has less effect on robust networks.

In section 3.5, I discussed low network sensitivity (i.e., the average probability that the change of one input node will affect each of the nodes that it influences). Leclerc criticized others for arguing that high-K consistent networks are robust only on the basis of their reduced sensitivity. High-K networks also have an increased number of mutation-prone connections. Leclerc argues that if we assume a constant mutation rate per regulatory connection, the increased number of connections overwhelms the reduced sensitivity to result in higher-K gene regulation networks' being less robust than lower-K gene regulation networks.

Under stabilizing selection, robustness is favored by natural selection. Stabilizing selection is a common assumption in evolutionary modeling. In Leclerc's models, the optimum was fixed and stable selection favored well-functioning lower-K networks. This result suggests a selective pressure in favor of low-K gene regulation networks. Leclerc believes that the most adaptive regulation of a gene will typically involve multiple transcription factors because of the variety of optimal gene expression profiles across genes and the relatively small number of transcription factors. (See section 4.4 above.) He sees the adaptivity of high-K networks and the robustness of low-K networks as two competing selective pressures on K in gene regulation networks. Natural selection typically favors the most adaptive tradeoff between selection pressures. For example, in antelopes natural selection favors strong legs to avoid injury and light legs for speed, but lighter legs tend to be weaker. The result is evolution of legs that are quite light but strong enough that they hold up to the typical rigors of escaping predators. Leclerc describes the case of gene regulation as follows:

> This indicates that sparse networks are actually more robust if the costs of complexity are accounted for. If true, then evolution should seek to optimize the costs and

benefits of complexity with a parsimonious network structure, a network topology that is sparsely connected and not unnecessarily complex, by seeking an optimal topological ensemble of interactions that best meets the network's functional requirements under its normal range of operating conditions. (2008, p. 3)

The selective pressures for accuracy and robustness work on different units of selection. Network accuracy is a property of an individual, so natural selection may favor an individual with a more accurate gene regulation network over another individual with a less accurate network. Robustness to mutation is not a property of an individual. Mutations happen across generations, so robustness to mutation is expressed and selected for only across generations. It is a property of a lineage of at least parent and offspring, but one might think of lineages with many more generations too. Natural selection may favor one lineage over another because it is more robust. The failure of an individual to reproduce is selection against that individual and that individual's lineage. (For a detailed discussion of lineage selection, see Sansom 2007.)

In section 4.5, I argued that increasing K in gene regulation networks improves evolvability. This is another selection pressure on K. Evolvability is the potential for mutations to be adaptive. It is a feature of mutation, and so it can be expressed only by lineages and can be selected for only on lineages. In this respect it is just like robustness to mutation, but the selective pressure for evolvability favors high K whereas the selection pressure for robustness favors low K.

It will be hard to figure out which selection pressure on lineages is stronger. Leclerc is encouraged about the importance of natural selection for low K by some empirical data. I gave reasons for caution in concluding that gene regulation networks are low K from these data in section 3.4. However, even as we learn more about K in complex organisms, we cannot conclude that this is the level of complexity at which lineage selection for robustness and lineage selection for evolvability have reached equilibrium, because individual selection for accuracy is also in play. Consider the following scenario as an example: Networks have evolved to be high-K because only high-K networks can be extremely accurate. This is the result of selection for individuals with accurate networks. Natural selection favors low K for robustness at the lineage level because the cost of the increased number of mutations in lineages with high-K individuals is greater than the evolvability benefit of those mutations tending to have smaller effects and thereby improving evolvability. This is a scenario whereby selection at the individual level for high K is keeping selection at the lineage level out of equilibrium.

Investigation into this issue with simulations will also be difficult. In Leclerc's simulations, low-K networks were able to perform well. Learning more about the tradeoff between lineage selection for robustness and individual selection for accuracy will require that the task for the network be difficult enough so that low-K networks cannot be as accurate as high-K networks. Learning more about selection between lineage selection for robustness and lineage selection for evolvability probably will require giving up the assumption of stabilizing selection. Under stabilizing selection, a population can reach a point at which there is no need to evolve or no ability to evolve more adaptively and thus no selective pressure on lineages for evolvability.

Whenever one identifies a selective pressure for a trait value, it is always worthwhile to think about other selective pressures on that trait value. Accordingly, it is worth keeping robustness to mutation in mind when considering evolvability.

4.7 An Evolutionary Trend in Gene Regulation Toward Greater Complexity

In sections 4.4 and 4.5, I suggested that because gene regulation networks are qualitatively consistent, increasing the number of transcription factors regulating each structural gene offers the opportunity for increased network accuracy and increased evolvability. Given a gene regulation network of reasonable complexity, there will always be a supply of additional transcription factors to add to the regulation of a particular structural gene. In this section, I explain how the way this supply can increase might result in an evolutionary trend toward greater complexity.

Thinking about gene regulation networks as connectionist networks allows certain hypotheses to seem obvious. For example, organisms with more complex development will have more complex gene regulation networks. In complex organisms, the network must distinguish between more microenvironments in which the adaptive genes to express are different. This adaptive complexity can be increased in a number of ways. Increasing the number of genes that produce transcription factors makes additional potentially adaptive regulatory connections possible.

Mutation to some features of development will have wider effects than mutations to other features. Such components are more *generatively entrenched* (Wimsatt 1986). Generally, organs that appear earlier in development and are necessary to usual development of more subsequent structures are more generatively entrenched than mutations to organs that appear later in development. Accordingly, their evolution is expected to be more conserved than that of less generatively entrenched features. This explains why humans and chickens look quite similar at early stages of development.

Consideration of the generative entrenchment of transcription factors suggests the possibility of a trend toward greater complexity in gene regulation networks. The regulatory and transcribed regions of transcription factor X may be duplicated in mutation to produce X'. In subsequent generations, X' may continue to perform the regulatory duties for which X has a long selective history. X may also influence those target genes. However, after duplication, mutation to X will have fewer effects on target genes than it would have had before the duplication. Consider a mutation that changed the regulatory region of X such that it is no longer expressed in a range of microenvironments R. Without the duplicate, this would have resulted in significant change in the regulation of the target genes in R, but with X' still expressed in R there is no change. Additionally, mutations to the transcribed region of X that remove or modify regulatory relationships with old target genes may have no effect on target gene expression in microenvironments in which X' is expressed. After mutations to the transcribed region of X, mutations to some of the original target genes might result in them being regulated by X but not X'. In such a scenario, both X and X' have fewer regulatory responsibilities than X did before the duplication, so both transcriptions factors are less generatively entrenched and more evolvable than X was.

An example that is consistent with the scenario laid out above is provided by the duplicate genes *Eng1* and *Eng2* in zebrafish (Force et al. 2004). This duplication occurred after ray-finned fishes diverged from tetrapods, which have only the precursor *En1*. In tetrapods, *En1* is expressed and functional in the pectoral appendage bud and in hindbrain and spinal cord interneurons. In zebrafish, *Eng1* is functionally expressed only in the pectoral appendage bud, and *Eng2* is only expressed in hindbrain and spinal cord interneurons. Multiple studies show this pattern of duplicate genes' performing complimentary subsets of the functions of the ancestral genes (Westin and Lardelli 1997; Nomes et al. 1998; Force et al. 1999; De Martino et al. 2000; Lister and Riable 2001).

A newly created duplicate transcription factor will not be highly generatively entrenched, because of the presence of the copy. It will add evolvability. It will also reduce robustness, because it will add another gene that may potentially mutate (see section 4.6), but that reduction in robustness will be greatly reduced by the presence of its duplicate. As the duplicate pair differentiates by mutation, each of the genes probably will take on overlapping subsets of the duties of its ancestor gene. As fewer of these duties become shared, each gene will become increasingly generatively entrenched. Increasing the number of transcription factors by duplication appears to greatly increase evolvability with limited reduction in robustness. In contrast, simply eliminating a functional transcription factor with no duplicate probably will be highly maladaptive, because transcription factors without duplicates are generatively entrenched. Even though each signal from one transcription factor may be substituted by another with no loss in adaptivity, the probability that all the mutations to target gene regulatory regions will occur in order for this to happen is low. On the assumption that producing increasing numbers of transcription factors is not overly costly in terms of the resources needed to produce the proteins themselves, we have uncovered an explanation of a ratcheting up of complexity at the level of gene regulation. If adding duplicate transcription factors is a less costly way to evolve than removing generatively entrenched transcription factors, then the need to adapt to a changing environment may result in a directional trend toward increasing complexity in the gene regulation network. This trend in trait value would not rely on any feature of the environment other than the fact that it changes.

4.8 Conclusion

I showed in chapter 2 that the evolvability of gene regulation networks is contingent on the fitness landscape of variants. There I described Kauffman's argument that gene regulation networks were not highly evolvable, which was based on a model with a fitness landscape dominated by large plateaus of unfit variants (figure 2.2). Such a fitness landscape is not conducive to adaptive evolution by natural selection. I offered an alternative model with a smoother fitness landscape of steadily increasing fitness (figure 2.3). In this chapter, I have argued that gene regulation networks of high *K* are highly evolvable, and that their evolvability can be understood using the connectionist framework. This leads naturally to the question What does the fitness landscape of connectionist networks look like?

The adaptivity of a connectionist network is determined largely by the accuracy of its indicators. Given the variety of expression profiles of transcription factors, we expect that the accuracy of potential indicators for a gene regulation network to be distributed in a way approximating a normal distribution (although it may be positively or negatively skewed). Highly accurate X indicators (always and only present in microenvironments in which the transcription of gene X is adaptive) are unusual, as are highly inaccurate X indicators (although the latter are accurate not-X indicators). The accuracy of networks of random collections of indicators is distributed normally, but with a smaller standard deviation than the distribution of indicators, because the probability of randomly selecting a network made up of only highly accurate indicators is less than that of picking just one accurate indicator. Nevertheless, if we assume an approximately normal distribution of indicator accuracy, the shape of the fitness landscape of connectionist networks approximates the fitness landscape of the more evolvable model that I introduced in chapter 2, which is itself a normal distribution. In particular, it allows gradual mutation as more accurate or less accurate indicators are added, each mutation having a decent chance of slightly altering the network's accuracy, rather than the vast majority being fitness neutral, as in the case of Kauffman's less evolvable model (discussed in chapter 2). This is another simpler way to see why gene regulation networks are evolvable according to the connectionist framework.

The qualitative consistency of regulatory connections between transcription factors and target genes is necessary if the connectionist framework is to be applicable to gene regulation. Biologists have had evidence that gene regulation is typically qualitatively consistent for a while now. Ultimately, the connectionist framework for gene regulation shows the importance of qualitative consistency to the adaptivity and evolvability of gene regulation networks.

5 Why Gene Regulation Networks Are the Controllers of Development

Consider a scene that took place on the banks of the Hydaspes River in the Punjab region of India in 326 BC. Nearly 50,000 infantry, more than 10,000 cavalry, and 200 elephants were involved in intense activity that resulted in more than 15,000 deaths. In order to make any sense of what occurred, it is necessary to understand who was controlling what was going on. It turns out that the forces that began on the right bank of the river, under the command of Alexander the Great, were battling the forces of the Punjabi King Porus, which were lined up on the left bank to repel any crossing. In response, Alexander had his forces carry out a pincer maneuver. Alexander led a small group of his forces 17 miles upstream. If Porus sent too many of his men upstream, Alexander's General Craterus would lead the bulk of his army across the river. Instead, Porus sent too few men to repel Alexander's crossing, and they were overwhelmed. Alexander was then able to bring all of his forces across the river. When the two complete forces later met, the Macedonian forces were ultimately able to surround the Indian forces after the Macedonian cavalry outflanked and circled behind the Indian lines. The great achievement of this maneuver was that it minimized the exposure of the Macedonian cavalry to the Indian war elephants, which was important because just the smell of the elephants scared the horses (Fuller 1960; Plutarch 75). Understanding such a battle requires identifying who was in control and what that person was trying to do. Without that information, we would see only a whirling confusion.

Biological development, particularly the development of complex organisms, is an enormously complex dynamic process. In order to understand such a system, it is crucial to understand the controller of that system. Just as understanding control offers insight into the activity that took place on an ancient battlefield, so too can understanding control offer insight into the activity that is biological development.

Searching for the controller of development requires investigating both the notion of control and the process and evolution of development. Different types of systems fall under different types of control. Military control is relatively clear and well understood. In many uniforms, the crucial information about control relationships is literally worn on one's sleeve. In this chapter, I will investigate various types of control with the goal of understanding the control that is exhibited in development. I will identify a new extrinsic concept, *design-control,* and suggest that gene regulation networks are the controllers of development. This argument will rely on the empirical fact that gene regulation networks are highly adaptive evolved systems. In chapter 4, I provided an explanation for the adaptivity and evolvability of gene regulation networks that rested on the connectionist framework. Thus, I will argue that the connectionist framework for gene regulation networks explains why gene regulation networks are the controllers of development, and why understanding gene regulation networks is crucial to understanding development.

5.1 Causal Control

Philosophers typically discuss the concept of control as it pertains to actions within the context of action theory. For example, in her investigation of how to apply reasons to animals, Susan Hurley (2001, p. 424) defines control as "the maintenance of a target value by endogenous adjustments in the context of exogenous disturbances." I shall begin my investigation of control by analyzing this definition.

To say that system *C* is under control relative to target value *T* of variable *V* is to say something about the process involving the system *C* and its environment that determines the value of *V*. In particular, changes in *V* from *T* due to changes outside system *C* are counteracted by changes within system *C* that return the value of *V* to *T*. I call Hurley's notion of control *causal control.* The notion of causal control is metaphysically lean. It places no restriction on what counts as a target value to be maintained, as an endogenous change, or as an exogenous disturbance, or on how the system and the environment are distinguished. This lack of conditions leaves the concept simple and flexible. The only metaphysical extravagance comes in the holistic nature of the concept, because it concerns how the whole system causally interacts with its environment. Causal control is determined only by the system and its environment's occurrent modal properties.

Causal control is applicable to systems under any type of control. For example, the temperature of an air-conditioned house is under control because, despite external heating of the house by the sun, the temperature (V) is maintained at the target value at which the thermostat is set (T) by the process of the air-conditioner's being turned on and off by the thermostat. Another example is that Alexander's forces are under Alexander's control. In this case, the value T that is maintained is the congruence between Alexander's intentions and the actions of his forces.

Interpretations of the causal control of a system are based on the division between a system and its environment, which determines what counts as an endogenous change rather than as an exogenous change within a process. Systems of any complexity can be conceptually divided into subsystems. Accordingly, an interpretation of control that involves a system^1 in its environment^1 can be consistent with another interpretation that identifies one of its subsystems as the system (system^2) and parts of system^1 as part of system^2's environment^2. Further nested interpretations may be possible (in which system^3 is a part of system^2) and alternative interpretations (in which system^4 is a part of system^1 but not a part of system^2).

For example, we can see the house and its thermostatically controlled air-conditioner as a system under control with respect to temperature, because the house and the air-conditioner maintain temperature T in response to external heating. One can also offer a nested interpretation in which only the thermostat is a system under control because it responds to the temperature in the house to maintain a value of turning the air conditioner on only if the temperature is above T. An alternative interpretation is to take the air conditioner by itself to be the system under control because it maintains a value of being on only if it is turned on by the thermostat, thereby maintaining T.

Dividing a system into a combination of subsystems seems reasonable. A system need not be physically contiguous. The human immune system is a non-contiguous system, for example. Though there should be a good deal of freedom to divide a whole system into different combinations of subsystems, we should always have a reason for grouping things together into one system. (See Kauffman 1970.) Generally, I think it is good practice to follow the lead of empirical science in carving the world up into different systems.

Ultimately, I am searching for a concept that will make it practicable to identify the controller of an organism unambiguously. Causal control is too flexible a concept to do this alone, but it does provide a necessary condition for a subsystem's being the controller of a system. A controller of a process of a system must itself be a subsystem under control within that process. For example, an air-conditioning unit and a thermostat may be said to be a controller of the temperature of the house. The process includes the thermostat's responding to temperature, the air-conditioning system's responding to the thermostat, and the temperature's responding to the air-conditioning system. The air-conditioning unit and the thermostat make up a *control system*, and the rest of the house (i.e., its environment) can be understood as the *controlled system*. The target value is the range of house temperatures that corresponds to the thermostat settings for turning the air-conditioning unit on and off. The disturbance (exogenous to the control system) may be an increase in house temperature above the target value; the endogenous adjustment may be the air-conditioning unit's turning on. After the house has been cooled to the point at which the air-conditioning unit is shut off by the thermostat, that drop in temperature becomes the exogenous disturbance, causing the endogenous adjustment of the air-conditioning unit's turning off. Thus, the thermostat and the air conditioner constitute a candidate for the controller of the house temperature. As I pointed out earlier in this section, such causal control analyses are consistent with nested analyses, such as that in which the thermostat alone is the control system and the air conditioner and the room are the controlled system. Similarly, one could see Alexander and his officers, rather than Alexander alone, as the controller of Alexander's forces. In the next section, I will apply causal control to a relatively simple fraction of the activity of development (the regulation of the metabolism of lactose involving the *lac* operon) to show that this notion will not identify the controller of even that system, ultimately for the same reason.

5.2 Causal Control of the *lac* Operon

The first process of transcription regulation to be understood concerned the *lac* operon (Jacob and Monod 1961). *lacZYA* are the genes of lactose transportation and metabolism, arranged in an operon in *E. coli*. The repressor (*LacI*) is present in an unmodified state when lactose is not present in the cell, repressing the transcription of *lacZYA* by binding to the regulatory region of the operon. Therefore, *lacZYA* is not expressed in significant levels when lactose is not present in the cell. The presence of lactose results in the presence of a lactose metabolite *allolactose*, which binds to the protein *LacI* and modifies it, preventing repression. That is not the only influence on *lacZYA*, though. The prevention of repression is not sufficient for transcription at significant levels. *lacZYA* transcription must be induced, too. When *CAP* is bound to *cAMP*, it binds to the *lac* regulatory region to encourage transcription of *lacZYA*. And *cAMP* is available only when glucose is not present in the cell. The result of this is that the proteins necessary for lactose metabolism are transcribed only in the presence of lactose and the absence of glucose. This is adaptive because glucose is a more efficient source of energy than lactose.

One can analyze the aforementioned process as one in which *LacI*, *CAP*, *cAMP*, and the regulatory region of the *lac* operon form a control system within *E. coli*, maintaining the target adaptive transcription level of *lacZYA*. High levels of *LacZYA* are adaptive only in the presence of lactose and the absence of glucose, so the adaptive level of *LacZYA* is determined by the presence of these sugars. The protein *LacI* binds to the regulatory region (an endogenous adjustment to the control system), thereby repressing the expression of *lacZYA* in response to the absence of lactose (an exogenous disturbance), and *CAP* binds with *cAMP* and the regulatory region (another endogenous change) in the absence of glucose (another exogenous disturbance) to allow the expression of *lacZYA* only in the presence of lactose and the absence of glucose (maintaining the target value).

One can also analyze this process as one in which the transcription level of *lacZYA* controls the metabolism rate of lactose. There are different control analyses of the metabolism for the same reason that there are different control analyses for the thermostatically controlled house temperature. In both cases there is a negative feedback loop, and, according to the definition of causal control, any link in a negative feedback loop can be identified as controlling the rest of the process.

The ambiguity of applying the notion of control to the metabolism of lactose above is typical in development, because many parts of the whole system that is an organism are reacting to other parts. This leaves us free to see processes in development as complex systems of parts mutually controlling each other. Seeing an organism as a system of parts mutually controlling each other is an important insight, but it does not allow us to identify *the* controller of the development. That is something that the notion of causal control cannot do. Below I shall develop a more demanding notion of *design-control* in order to ultimately be able to identify *the* controller of a system.

5.3 Design-Control

Each of the different causal control analyses of the air-conditioned house and lactose metabolism offered above seem appropriate. The reaction to the following control analysis of the air-conditioned house may be different. We can analyze the insulation of a house as controlling how much the air-conditioning unit is on. Poorly insulated walls, for example, allow heat from the sun (an exogenous disturbance) to cause the temperature of the house to rise (an endogenous adjustment), maintaining a target value (the proportion of time that the air-conditioning unit is on). This interpretation relies on what some might take to be a peculiar target value (i.e., the proportion of time that the air-conditioning unit is on). We can make a similarly peculiar control analysis of lactose metabolism, according to which glucose and lactose levels control the expression of *lacZYA*. This too seems to put the controlled cart before the controlling horse.

These analyses are further evidence that the notion of causal control is very permissive. In this section, I will offer a different notion that excludes the peculiar control analyses just offered. My intention is to offer a notion of control—*design-control*—that narrows down the appropriate control analysis of a system as much as is possible.

Consider again the control analysis of the air-conditioned house that had the insulation of the house controlling the amount of time that the air-conditioning unit was on. What is wrong with this analysis? One might be inclined to think that something like the insulation of a house cannot be the controller of anything, because it is not sufficiently dynamic. The house is as insulated as it is. The presence of the insulation has a number of causal consequences, but it is not in any way reacting to anything. Contrast this with a thermostat controlling an air-conditioning unit, for example. It changes its output of turning the air-conditioning unit on or off in a more dynamic way (with a switch, no less). Though I appreciate how intuitive this reaction may be, I am not convinced that the distinction on which it relies is tenable. A house's warming as a result of poor insulation is a somewhat dynamic process because the temperature does change. Is it insufficiently dynamic because it is too predictable, or too regular? Again, I do not think that is right. If the air-conditioning unit was hooked up to a timer and ran only from 10 a.m. to 10 p.m., its behavior would be extremely regular, but I still think it would be controlled by the timer.

I think that the answer lies in the design of the components of the whole system. Thermostats are designed to react to changes in temperature. Insulation is designed to keep the temperature on one side of a wall different from the temperature on the other side and not to keep air-conditioning units on a certain amount of the time. To the extent that the house warms during a hot summer, the insulation fails at its function, although in a predictable and understandable way, or so we typically think. Such a notion of control that allows the thermostat to control the air-conditioning unit, but not the insulation, is a functional notion, which is determined by its history. The thermostat controls the air-conditioning unit because the whole system was designed by someone with that intention. This fact about the design of the whole system is historical. I call this notion *design-control*.

Once we see design-control as an historical property, we can even see that our initially implausible control interpretation of the air-conditioned house can be made plausible under the right design scenario. Consider an architect who knows the size of the house she wishes to design, the temperature above which she does not want the house to rise, the outside temperatures and solar radiation that can be expected over a year, the capacity of the air-conditioning system, and the air-conditioning system's power supply (a solar panel that will generate a certain amount of energy, and batteries that will store it). One might imagine the architect calculating the exact minimum insulation required for this whole system to maintain that target value temperature. In such a case, the function of the insulation is to allow a certain amount of heating but no more, and we might reasonably say that the insulation is a design-controller of the air-conditioning unit.

5.4 *The* Controller of a Process

Above, I described the notion of design-control and showed why it is more demanding and therefore more exclusive than casual control. Design-control is causal control of a target value by design. Some systems that qualify as causal controllers of a system do not qualify as design-controllers of that system (e.g. the insulation of the air-conditioned house as originally described). However, multiple systems might still design-control a target value. Thermostats are designed to activate air-conditioning units, and air-conditioning units are designed to cool houses when activated.

I think the notion of design-control also has the potential to rank the degree to which different systems design-control a particular process. One of the systems that design-controls a target value is the system that was most specifically designed to control that target value. That system is *the* controller of that process. In the example of the air-conditioned house that was designed for a limited heating system, the case for the insulation's being the design-controller rests on the fact that it was most specifically designed for that function. A great deal of effort may have gone into the design of the air-conditioning system, which required the greatest variety of components working together in a precise way for it to work at all. But this design was probably done completely independent of the application of cooling this particular house. The insulation level, in contrast, was designed specifically for this application. Curiously, we might say that the architect controls how much the air-conditioning unit will be on by adjusting the degree of insulation in the design. This historical fact makes the insulation *the* controller of the process. If a window is carelessly left open, the air-conditioning system will not keep the house below the specified temperature. The system will be under the causal control of the open window, the thermostat, the air-conditioning unit, and even the insulation, but it will not be under the design-control of the insulation. In fact, it will not be under the design-control of anything, because the whole system will not be acting as it was designed.

Above I described how a number of systems are involved in lactose metabolism in *E. coli*. Which of these systems can be identified as *the* controller of the process? According to my notion of design-control, this depends on the details of the evolutionary history of *E. coli*. It depends on which system has been most precisely designed by natural selection to carry out its controlling role.

Some may object to applying the notion of design to evolution by natural selection. Certainly there are differences between the process that lead to *E. coli*'s being capable of metabolizing lactose and the architect's designing the cooling system of the house. In particular, natural selection never anticipated a possible outcome and then tried to bring it about. Instead, natural selection only favored trait values for the specific ways that they were adaptive. If applying the term design-control to products of evolution is too much for some, they may prefer a more neutral term (perhaps *functional control*), but I think it is appropriate to apply the notion to design-control to the products of evolution by natural selection as long as the differences between evolution by natural selection and intentional design are kept in mind.

LacI is a transcription factor and *LacZYA* are structural proteins influenced by that transcription factor. The relationship between *LacI* and *LacZYA* is due to the structure of the *LacI* protein and the regulatory region of *lacZYA*. The effect of *LacZYA* on the metabolism of lactose is due to the transcribed region of *lacZYA*. We may boil down the question of whether the metabolism of lactose is controlled by the protein *LacI*, by the proteins *LacZYA*, or by the relationship between them to the question of which of these three things has been most specifically designed for this particular application. The details of the evolution of transcription factors, structural proteins, and the relationship between them is not known in most cases, but some general lessons are emerging. I shall return to this issue in the next section.

5.5 *The* Controller of a System

My ultimate goal in this chapter is to narrow *the* controller of development down to one system in order to see what insight this offers into development. So far, I have discussed only the design-control of a process that a whole system carries out. Now I must show the characteristics of the controller of a system simpliciter. The controller of a system simpliciter is *the* controller of the widest range of processes performed by a system, which is the system that was most specifically designed to control each of the widest range of processes performed by a system.

Use of this criterion requires that processes performed by a whole system be distinguished from one another. The issue of distinguishing between processes is largely the same as the issue (discussed above) of distinguishing between systems, because systems are processes, and my approach is the same. It seems reasonable to have the freedom to distinguish processes in more than one way, but each process should have some unifying feature, and it is good practice to respect the divisions made by empirical science.

If the controller of a whole system is the controller of the widest range of processes, and the design-controller of a process is the system most specifically designed for each process, then, all things being equal, we would expect the controller of a whole system to be the most designed system within the whole system. This is not a deductive argument, but a reasonable rule of thumb. Let me summarize the relationship between this rule of thumb about the controller of a system and the definition of the design-controller of a process. By definition, the most specifically designed system that causally controls a process is the design-controller of that process. By definition, *the* controller of a whole system is the design-controller of the widest range of processes carried out by the whole system. Figuring out which system satisfies this definition in a complex system is no mean feat, so I suggest the rule of thumb that *the* controller of a whole system is *likely* to be the most designed system in that whole system. There may be exceptions to this rule of thumb, because it is not a definition, but it may still prove useful.

5.6 *The* Controller of Development

If we apply the rule of thumb to development, then the controller of development is likely to be the system most designed by natural selection. That system will have the most adaptive mutations in its history that were effectively selected within historical populations by natural selection. A single mutation may involve changes to many systems of an organism—for example, a change in DNA sequence, a potential change in proteome or gene regulation, a change in some kind of cellular activity, a change in tissues, organs, a change in organism behavior, and so on. At which level should we expect to find the controller of development?

All things being equal, we should expect to find the controller of development at a fairly low level because much of development takes place before high levels (e.g. the organ level) are developed. This should not be the only consideration. Systems not included in a zygote are not thereby disqualified from consideration. However, any development for which a system cannot be responsible speaks against that system's being the controller of development.

To find the most designed system, we have two criteria that seem very much at odds. First, the controlling system should have a higher number of adaptive mutations in its history than any other system. The more broadly we conceive of the controlling system, the more adaptive mutations there are in its history. Second, in order to provide the most insight into the operations of the whole system, the controlling system should be well specified (i.e., as much of the whole system should be excluded from the controlling system as is possible). For example, to propose that the whole organism is its own controller is to give an answer that fully satisfies the first criterion but completely fails to satisfy the second.

A vast majority of heritable mutations involve a change in DNA sequence, but the goal of specificity requires that we at least consider distinguishing the evolution of regulatory regions from the evolution of transcribed regions. As was discussed in the previous chapter, variation in regulatory regions has been linked to a wide range of phenotypic variations. Additionally, a relatively low proportion of active sites on regulatory regions are common across species. For example, a survey of twenty regulatory regions in mammals showed that probably only two-thirds of binding sites in humans are functional in rodents (Dermitzakis et al. 2002). This finding is an example of the growing collection of evidence that more adaptive evolution takes place in regulatory regions than in transcribed regions (Gerhart and Kirschner 1997; Stern 2000; Carroll et al. 2001; Wilkins 2002). Both regions of a structural gene may have evolved to carry out their functions, but in general the regulatory region is where the fine tuning takes place, as the high degree of adaptively significant variation between species attests. Though the proteins *LacZYA* may have evolved to metabolize lactose, the details of just to what degree *LacI* represses their expression in *E. coli* probably have been more finely evolved. As Lenny Moss puts it, development has evolved not by changing the script of development, but by changing the nature of the "ad hoc committees" that determine transcription, splicing, and so on (2003, p. 189).

Regulatory regions of structural genes, together with the regulatory and transcribed regions of transcription factors, determine the gene regulation network of an organism. Because regulatory regions appear to be so highly evolved, I propose that gene regulation networks are the most designed system in an organism. The rule of thumb about the controller of a system, described above, suggests that gene regulation networks are the controllers of development. How well do gene regulation networks meet the three criteria that determine which system is the controller of the whole system? Gene regulation networks play a role in all processes that involve proteins transcribed from DNA, so their influence is ubiquitous throughout development. The gene regulation network is functional in the zygote. The regulatory regions are fully formed. Most transcription factors are yet to be produced, but their transcribed regions are determined. In short, much of the gene regulation network is working early in development. There are some processes, such as the firing of nerves and the contractions of muscles, that are not under the direct control of the gene regulation network (although these processes do require structures that develop under the control of gene regulation networks). It is worth investigating whether control might shift from one system to another during development. All things considered, I propose that development as a whole is controlled by gene regulation networks, because they are finely evolved by natural selection, they are involved in the earliest development, and they continue to be crucial throughout development.

5.7 Skepticism About Gene-centrism

I have argued that gene regulation networks are the controllers of development. This view is likely to be met with initial skepticism among biology theorists, because many views about the importance of genes have not stood up well to close inspection. In chapter 1, I discussed perhaps the most famous gene-centric view of evolution: Richard Dawkins's (1976) view that genes are the fundamental unit of selection. Another attempt was John Maynard Smith's (1993) view that only genes have the function of carrying information. These and other views attempted to draw a conceptual qualitative difference between genes and other developmental resources and failed in nearly every case because developmental resources other than genes meet the same standard. Organisms and groups, for example, satisfy the criteria for being units of selection, and environmental factors (such as nest sites, birdsongs, and other heritable parenting techniques) all have the evolutionary histories and the causal influence necessary for having the function of bearing information. Indeed, one of the calling cards for an entire view of evolutionary theory (developmental systems theory) is the "parity" thesis, which holds that a distinction between genes and everything else should not lie at the heart of evolutionary theory (Sterelny and Griffiths 1999).

There are two reasons why the position defended here is different from those presented above. The first is that my thesis concerns the gene regulation network, not the gene. This minor difference will not impress many, because there is a sense in which gene regulation is just one step removed from the genes themselves. The second and more important reason is that I have not argued for a conceptual difference between gene regulation networks and everything else. Rather, I have argued for the validity of a measure for how much a system controls development. This measure allows different systems to be compared in terms of the degree of design-control that they have. Accordingly, it allows one system to be distinguished as the system that exhibits the most design-control, and that system is said to be *the* controller of development. The scale of design-control is measured in terms of the number of adaptations specific to the task currently performed by the system. Genes and many epigenetic systems of inheritance have adaptive mutations, and I draw no qualitative distinctions between their adaptations and those of gene regulation networks. Rather, I offer a quantitative distinction between these systems and gene regulation networks in terms of number of adaptations.

Some may object that, insofar as I fundamentally offer only a quantitative distinction, I should not talk about *the* controller at all, but should simply accept that design-control is shared in various degrees across systems. That is a valid point of view and should not be lost in any discussion of the controller of development. However, I think that identifying the controller offers insight into many different types of systems—from developing organisms to and battling military forces, to social democracies. In some cases, such as military forces, identifying the controller may be straightforward. In other cases, such as a social democracy, control is shared, and whether the population is the controller of their leadership or vice versa may be a difficult question to answer. Leaders and populations may influence each other, but which has more control? This is an important issue for our understanding of any political system, and it may turn on slight differences regarding such factors as integrity of legal institutions, freedom of the press, or the attitudes of leaders of the military. In general, I expect that there is a fact of the matter, and even if that fact is hard to obtain, it is worth the effort. I feel that the same is true about the controller of development.

5.8 Ingenious Genes

This book began with the question of how a mutation can lead to adaptive complexity in an organism. I suggest that an important step toward answering that question is to apply the connectionist framework to gene regulation networks. Doing so suggests that we see an organism as a collection of cells, each cell controlled by the same gene regulation network to modify its activity on the basis of its particular microenvironment. The gene regulation network ensures that each cell does the right thing at the right time, responding to various features in the microenvironment, including levels of nutrients, hormones, and signals from other cells, in a way that is adaptive for the organism. To the extent that we have an intuitive handle on connectionist networks in general, we have an intuitive answer to a crucial aspect of Waddington's question: How does an organism get up and start walking about?

There is growing interest in the relationship between evolution and development. The connectionist framework for gene regulation essentially integrates these two processes, because connectionism is not only a theory of how a well-designed network operates but also a theory of how a network becomes well designed. Accordingly, the connectionist framework offers insight not only into how a well-designed gene regulation network functions but also into how networks can evolve as regulatory relationships are added, removed, and modified. Gene regulation networks are highly evolvable in part because the relationships between transcription factors and genes are arbitrary. Any transcription factor protein may play a role in regulating any target gene with the right mutations to the target gene's regulatory region. Additionally, the qualitative consistency of regulatory relationships makes the addition of a novel relationship quite likely to be adaptive, because adding the influence of just one transcription factor on the expression of a gene probably will only have only a moderate effect on the expression profile of that gene.

The connectionist framework for gene regulation does not give us a complete understanding of development and its evolution. In particular, it does not include an explanation of exactly why small changes in gene expression profiles can result in small changes in development, or any explanation of various systems of homeostasis that take place at levels above the cell. However, it is an important part of the explanation of how novelty that adds complexity has a chance to be adaptive, and therefore it offers insight into the whirling activity that is development and the epic drama that is the evolution of life.

These factors combine to suggest that gene regulation networks are the most designed system in development. A gene is an ingenious system that is well designed to solve a specific problem, but genes interacting make up the most ingenious system in an organism that solves the most challenging aspect of the difficulty of development: the difficulty of gene regulation. That is why I claim that ingenious genes evolve to control development.

Notes

Chapter 1

1. Another problem that pressured Darwin to adopt the inheritance of acquired characteristics had to do with the age of the Earth. Darwin's estimate that the Earth was several hundred million years old was based on geological evidence, which he thought was long enough to account for the evolution by natural selection of complex life. The physicist William Thomson (later Lord Kelvin) held that the Sun's energy was either chemical or due to gravity and, therefore, that the Sun was at most a few tens of millions of years old. As the Earth could not be older than the Sun, it must be no more than one-tenth as old as Darwin thought. This appeared to deny evolution the time necessary to evolve the complex life seen in Darwin's time by natural selection alone. That this objection underestimated the age of the Earth (in part because it did not take into account radiation as a source of energy) was not discovered until after Darwin's death.

Chapter 2

1. For examples of the variance of gene expression across life cycle, cell type, physiological condition and environmental condition, see White et al. 1999; Iyer et al. 2001; Kayo et al. 2001; Arbeitman et al. 2002.

2. I have shown that simulations with a fitness landscape different from Kauffman's result in higher equilibrium adaptiveness under selection. Jeffrey Schank and William Wimsatt (1988) have pointed out another variable that is ignored by Kauffman's fitness landscape that impacts its results. Wimsatt (1986) introduced the notion of "generative entrenchment," which includes the idea that certain traits have a greater impact on fitness than others. When some connections influence fitness more than others, natural selection does a better job of designing networks to have those connections.

Chapter 3

1. Readers will recall that this is the title of Kauffman's 1995 book.

2. This size network is of particular interest to Kauffman because he was working on the best assumption at the time, which was that humans have approximately 100,000 genes. The current view is that humans have 20,000–25,000 genes (International Human Genome Sequencing Consortium 2004). An attractor in a $K = 2$ Kauffman network with $N = 22{,}500$ has 150 states on average.

3. Some qualitatively inconsistent transcription factor exceptions and their importance to the model are discussed later in this section.

4. On the assumption that human gene regulation networks are generic $n = 100{,}000$ $K = 2$ networks, we would expect them to have attractors of 317 states, because 317 is the square root of 100,000.

5. Mark Changizi, personal correspondence.

6. I will discuss this example further in chapter 5.

Chapter 4

1. Condorcet's jury theorem assumed that each voter had an equal vote. This is not the appropriate assumption for quantitatively connectionist networks, in which some nodes have stronger connections than others. Nor is it appropriate for qualitatively connectionist networks, in which nodes can vary in connection strength with each context the network faces. The potential of varying weight strengths have been investigated in the political literature in the search for the optimal weighting for each voter if they vary in their accuracy (Nitzan and Paroush 1982; Shapley and Grofman 1984).

References

Abouheif, Ehab H., and Gregory A. Wray. 2002. The developmental genetic basis for the evolution of wing polyphenism in ants. *Science* 297: 249–252.

Abraham, Irene, and Winifred W. Doane. 1978. Genetic regulation of tissue-specific expression of amylase structural genes in *Drosophila melanogaster. Proceedings of the National Academy of Sciences USA* 75: 4446–4450.

Allendorf, Fred W., Kathy L. Knudsen, and Stevan R. Phelps. 1982. Identification of a gene regulating the tissue expression of a phosphoglucomutase locus in rainbow trout. *Genetics* 102: 259–268.

Allendorf, Fred W., Kathy L. Knudsen, and Robb F. Leary. 1983. Adaptive significance of differences in the tissue-specific expression of a phosphoglucomutase gene in rainbow trout. *Proceedings of the National Academy of Sciences USA* 80: 1397–1400.

Anisimov, Sergey V., Maria V. Volkova, Lilia V. Lenskaya, Vladimir K. Khavinson, Dina V. Solovieva, and Eugene I. Schwartz. 2001. Age-associated accumulation of the apolipoprotein C-III gene T-455C polymorphism C allele in a Russian population. *Journals of Gerontology* Series A, *Biological Sciences and Medical Sciences* 56: B27–B32.

Arbeitman, Michelle N., et al. 2002. Gene expression during life cycle of *Drosophila melanogaster*. *Science* 297: 2270–2275.

Arnone, M. I., and E. H. Davidson. 1997. The hardwiring of development: Organization and function of genomic regulation systems. *Development* 124: 1851–1864.

Arthur, W. 2000. The concept of developmental reprogramming and the quest for an inclusive theory of evolutionary mechanisms. *Evolution & Development* 2 (1): 49–57.

Averof, Michalis, and Nipam H. Patel. 1997. Crustacean appendage association with changes in *Hox* gene expression. *Nature* 388: 682–686.

Avery, Oswald T., Colin M. MacLeod, and Maclyn McCarty. 1944. Studies on the chemical nature of the substance inducing transformation of pneumococcal types. *Journal of Experimental Medicine* 79 (2): 137–158.

Barlow, Nora, ed. 1958. *The Autobiography of Charles Darwin 1809–1882*. Collins.

Bateson, William. 1914. The address of the president of the British Association for the Advancement of Science: Dr. William Bateson. *Science* New Series 40 (1026): 287–302.

Beadle, George, and Edward Tatum. 1941. Genetic control of biochemical reactions in Neurospora. *Proceedings of the National Academy of Sciences USA* 27: 499–506.

Behe, Michael. 1996. *Darwin's Black Box: The Biochemical Challenge to Evolution*. Free Press.

Bell, G., and A. O. Mooers. 1997. Size and complexity among multicellular organisms. *Biological Journal of the Linnean Society* 60: 345–363.

Benecke, A., C. Gaudon, and H. Gronemeyer. 2001. Transcription integration of hormone and metabolic signals by nuclear receptors. In *Transcription Factors*, ed. J. Locker. Academic Press.

Brakefield, Paul M., Julie Gates, Dave Keys, Fanja Kesbeke, Pieter J. Wijnaarden, Antóonia Monteiro, Vernon French, and Sean B. Carroll. 1996. Development, plasticity and evolution of butterfly eyespot patterns. *Nature* 384: 236–242.

Brylski, P., and B. K. Hall. 1988a. Ontogeny of a macroevolutionary phenotype: The external cheek pouches of geomyoid rodents. *Evolution* 43: 391–395.

Brylski, P., and B. K. Hall. 1988b. Epithelial behaviors and threshold effects in the development and evolution of internal and external cheek pouches in rodents. *Zeitschrift für Zoologische Systematic und Evolutionsforschung* 26: 144–154.

Burian, Richard. 1986. On integrating the study of evolution and of development. In *Integrating Scientific Disciplines*, ed. W. Bechtel. Martinus Nijhoff.

Burke, A. C., C. E. Nelson, B. A. Morgan, and C. Tabin. 1995. *Hox* genes and the evolution of vertebrate axial morphology. *Development* 121: 333–346.

Calhoun, Vincent, Angelike Stathopoulos, and Michael Levine. 2002. Promoter-proximal tethering elements regulate enhancer-promoter specificity in the *Drosophila* Antennapedia complex. *Proceedings of the National Academy of Sciences USA* 99: 9243–9247.

Carey, Michael, and Stephen T. Smale. 2000. *Transcriptional Regulation in Eukaryotes: Concepts, Strategies, and Techniques*. Cold Spring Harbor Laboratory Press.

Carroll, Sean B. 2000. Endless forms: The evolution of gene regulation and morphological diversity. *Cell* 101: 577–580.

Carroll, Sean B., Jennifer K. Grenier, and Scott D. Weatherbee. 2001. *From DNA to Diversity: Molecular Genetics and the Evolution of Animal Design*. Blackwell Science.

Caspi, Avshalom, Joseph McClay, Terrie E. Moffitt, Jonathan Mill, Judy Martin, Ian W. Craig, Alan Taylor, and Richie Poulton. 2002. Role of genotype in the cycle of violence in maltreated children. *Science* 297: 851–854.

Costanzo, M. C., et al. 2001. YPD™, PombePD™ and WormPD™: Model organism volumes of the BioKnowledge™ Library, an integrated resource for protein information. *Nucleic Acids Research* 29: 75–79.

Crick, Francis. 1958. Central dogma of molecular biology. *Nature* 227: 561–563.

Darwin, Charles R. 1844. Letter to Joseph Dalton Hooker, 11 January. Darwin Correspondence Project, letter 729. Available at http: //www.darwinproject.ac.uk.

Darwin, Charles R. 1851. *Living Cirripedia, A Monograph on the Sub-Class Cirripedia, with Figures of All the Species. The Lepadidæ; Or, Pedunculated Cirripedes*. Ray Society.

Darwin, Charles R. 1854. *Living Cirripedia, The Balanidæ, (or sessile cirripedes); the Verrucidæ*. Ray Society.

Darwin, Charles R. 1856. Letter to Joseph Dalton Hooker, 13 July Letter number 1924: The Darwin Correspondence Project. http: //www.darwinproject.ac.uk/home

Darwin, Charles R. 1859. *On the Origin of Species by Means of Natural Selection or the Preservation of Favoured Races in the Struggle for Life*. John Murray.

Darwin, Charles R. 1861. *On the Origin of Species by Means of Natural Selection, or The Preservation of Favoured Races in the Struggle for Life*, third edition. John Murray.

Darwin, Erasmus. 1801. *Zoonomia, or The Laws of Organic Life*, third edition. J. Johnson.

Darwin, Erasmus. 1818. *Zoonomia: or, the Laws of Organic Life*, fourth American edition. Edward Earle and William Brown.

Davidson, Donald. 1987. Knowing one's own mind. *Proceedings and Addresses of the American Philosophical Association* 60: 441–458.

Davidson, Eric H. 2001. *Genomic Regulatory Systems: Development and Evolution*. Academic Press.

Dawkins, Richard. 1976. *The Selfish Gene*. Oxford University Press.

Dawkins, Richard. 1989. The evolution of evolvability. In *Artificial Life: Synthesis and Simulation of Living Systems*, ed. G. Langton. Addison-Wesley.

Dawkins, Richard. 1996. *Climbing Mount Improbable*. Norton.

De Martino, S., et al. 2000. Expression of sox11 gene duplicates in zebrafish suggests the reciprocal loss of ancestral gene expression patterns in development. *Developmental Dynamics* 217: 279–292.

Dermitzakis, Emmanouil T., and Andrew G. Clark. 2002. Evolution of transcription factor binding sites in mammalian gene regulatory regions: Conservation and turnover. *Molecular Biology and Evolution* 19: 1114–1121.

DiLeone, Ralph J., Liane B. Russell, and David M. Kingsley. 1998. An extensive 3' regulatory region controls expression of *Bmp5* in specific anatomical structures of the mouse embryo. *Genetics* 148: 401–408.

Doebley, J., and L. Lukens. 1998. Transcription regulators and the evolution of plant form. *Plant Cell* 10: 1075–1082.

Dudareva, N., L. Cseke, V. M. Blanc, and E. Pichersky. 1996. Evolution of floral scent in *Clarkia*: Novel patterns of S-linalool synthase gene expression in the *C. breweri* flower. *Plant Cell* 8: 1137–1148.

Enard, Wolfgang, Molly Przeworski, Simon E. Fisher, Cecilia S. L. Lai, Victor Wiebe, Takashi Kitano, Anthony P. Monaco, and Svante Pääbo. 2002. Molecular evolution of FOXP2, a gene involved in speech and language. *Nature* 418: 869–872.

Fang, Shu, Aya Takahashi, and Chung-I Wu. 2002. A mutation in the promoter of desaturase 2 is correlated with sexual isolation between *Drosophila* behavioral races. *Genetics* 162: 781–784.

Fisher, Ronald. 1930. *The Genetical Theory of Natural Selection.* Oxford University Press.

Force, Allan, Michael Lynch, F. Bryan Pickett, Angel Amores, Yi-Lin Yan, and John Postlethwait. 1999. Preservation of duplicate genes by complementary, degenerative mutations. *Genetics* 151: 1531–1545.

Force, A. G., A. C. William, and F. B. Pickett. 2004. Informational accretion, gene duplication, and the mechanisms of genetic module parcellation. In *Modularity in Development and Evolution*, ed. G. Schlosser and G. Wagner. University of Chicago Press.

Fuller, John. 1960. *The Generalship of Alexander the Great.* Da Capo.

Gerhart, J., and M. Kirschner. 1997. *Cells, Embryos, and Evolution: Toward a Cellular and Developmental Understanding of Phenotypic Variation and Evolutionary Adaptability.* Blackwell Science.

Gould, Stephen Jay. 1989. *Wonderful Life: The Burgess Shale and the Nature of History.* Norton.

Gould, Stephen Jay. 1997. *Full House: The Spread of Excellence from Plato to Darwin.* Harmony.

Gould, Stephen Jay. 2001. The evolutionary definition of selective agency, validation of the theory of hierarchical selection, and fallacy of selfish gene. In *Thinking About Evolution*, ed. R. Singh et al. Cambridge University Press.

Grefenstette, John J. 1987. *Proceedings of the Second International Conference on Genetic Algorithms and Their Application*. Erlbaum.

Griffiths, P. E., and E. M. Neumann-Held. 1999. The many faces of the gene. *Bioscience* 49: 656–662.

Hoyle, Fred. 1983. *The Intelligent Universe*. Michael Joseph.

Hurley, Susan. 2001. Overintellectualizing the mind. *Philosophy and Phenomenological Research* 63 (2): 423–431.

International Human Genome Sequencing Consortium. 2004. Finishing the euchromatic sequence of the human genome. *Nature* 431: 931–945.

Iyer, Vishwanath R., Christine E. Horak, Charles S. Scafe, David Botstein, Michael Snyder, and Patrick O. Brown. 2001. Genomic binding sites of the yeast cell-cycle transcription factors SBF and MBF. *Nature* 409: 533–538.

Jacob, F., and J. Monod. 1961. On the regulation of gene activity. *Cold Spring Harbor Symposia on Quantitative Biology* 26: 193–211.

Kammandel, B., K. Chowdhury, A. Stoykova, S. Aparicio, S. Brenner, and P. Gruss. 1999. Distinct *cis*-essential modules direct the time-space pattern of the *Pax6* gene activity. *Developmental Biology* 205: 79–97.

Kauffman, Stuart A. 1970. Articulation of parts explanation in biology and the rational search for them. In *PSA 1970*, ed. R. Buck and R. Cohen. Reidel.

Kauffman, Stuart A. 1985. Self-organization, selective adaptation and its limits: A new pattern of inference in evolution and development. In *Evolution at a Crossroads: The New Biology and the New Philosophy of Science*, ed. D. Depew and B. Weber. MIT Press.

Kauffman, Stuart A. 1990. Requirements for evolvability in complex systems: Orderly dynamics and frozen components. *Physica D* 42 (1–3): 135–152.

Kauffman, Stuart A. 1993. *The Origins of Order: Self-organization and Selection in Evolution*. Oxford University Press.

Kauffman, Stuart A. 1995. *At Home in the Universe: The Search for Laws of Self-Organization and Complexity*. Oxford University Press.

Kauffman, S., C. Peterson, B. Samuelsson, and C. Troein. 2003. Random Boolean network models and the yeast transcriptional network. *Proceedings of the National Academy of Sciences USA* 100: 14796–14799.

Kayo, Tsuyoshi, David B. Allison, Richard Weindruch, and Thomas A. Prolla. 2001. Influences of aging and caloric restriction on the transcription profile of skeletal muscle from rhesus monkeys. *Proceedings of the National Academy of Sciences USA* 98: 5093–5098.

Kelley, K. M., H. Wang, and M. Ratnam. 2003. Dual regulation of ets-activated gene expression by SP1. *Gene* 307: 87–97.

Kirchhamer, Carmen V., Leonard D. Bogarad, and Eric H. Davidson. 1996. Developmental expression of synthetic cis-regulatory systems composed of spatial control elements from two different genes. *Proceedings of the National Academy of Sciences USA* 93: 13849–13854.

Laakso, Aarre, and Garrison Cottrell. 2000. Content and cluster analysis: Assessing representational similarity in neural systems. *Philosophical Psychology* 13 (1): 47–76.

Lamarck, Jean-Baptiste. 1797. *Memoires de physique et d'histoire naturelle*. Agasse.

Lamarck, Jean-Baptiste. 1801. *Système des Animaux sans vertèbres, ou Tableau général des classes, des ordres et des genres de ces animaux; présentant leurs caractères essentiels et leur distribution d'après les considérations de leurs rapports naturels et de leur organisation, et suivant l'arrangement établi dans les galeries du Muséum d'Histoire naturelle, parmi leurs dépouilles conservées; précédé du Discours d'ouverture de l'an VIII de la République*. Déterville.

Lamarck, Jean-Baptiste. 1802. *Recherches sur l'organisation des corps vivans et particulièrement sur son origine, sur la cause de ses développemens et des progrès de sa composition, et sur celle qui, tendant continuellement à la détruire dans chaque individu, amène nécessairement sa mort; précédé du discours d'ouverture du cours de zoologie, donné dans le Muséum national d'Histoire Naturelle*. Maillard.

Lamarck, Jean-Baptiste. 1815. *Histoire naturelle des animaux sans vertèbres, présentant les caractères généraux et particuliers de ces animaux, leur distribution, leurs classes, leurs familles, leurs genres, et la citation des principales espèces qui s'y rapportent; précédée d'une introduction offrant la détermination des caractères essentiels de l'animal, sa distinction du végétal et des autres corps naturels; enfin, l'exposition des principes fondamentaux de la zoologie*. Déterville.

Latchman, David S. 1998. *Eukaryyotic Transcription Factors*. Academic Press.

Laubichler, Manfred, and Jane Maienschein. 2007. Embryos, cells, genes, and organisms: A few reflections on the history of evolutionary developmental biology. In *Integrating Evolution and Development: From Theory to Practice*, ed. R. Sansom and R. Brandon. MIT Press.

Lee, Tong Ihn, et al. 2002. Transcriptional regulatory networks in *Saccharomyces cerevisiae*. *Science* 298: 799–804.

Lemon, Bryan, and Robert Tjian. 2000. Orchestrated response: A symphony of transcriptional factors for gene control. *Genes & Development* 14: 2551–2569.

Lerman, Daniel N., Pawal Michalak, Amanda Helin, Brian R. Bettencourt, and Martin E. Feder. 2003. Modification of heat-shock gene expression in *Drosophila melanogaster* populations via transposable elements. *Molecular Biology and Evolution* 20: 135–144.

Lettice, Laura A., et al. 2002. Disruption of a long-range *cis*-acting regulatory for Shh causes preaxial polydactyly. *Proceedings of the National Academy of Sciences USA* 99: 7548–7553.

Lewontin, R. C. 1978. Adaptation. *Scientific American* 239 (3): 156–169.

Lister, J. A., J. Close, and D. W. Raible. 2001. Duplicate mitf genes in zebrafish: Complementary expression and conservation of melanogenic potential. *Developmental Biology* 237: 333–334.

Locker, Joseph. 2001. *Transcription Factors*. Academic Press.

Lyell, Charles. 1830. *Principles of Geology*. John Murray.

Ma, S., Q. Gong, and H. Bohnert. 2007. An *Arabidopsis* gene network based on the graphical Gaussian model. *Genome Research* 17: 1614–1625.

Macdonald, Cynthia, and Graham Macdonald. 1995. *Connectionism: Debates on Psychological Explanation*, volume 2. Blackwell.

Malthus, Thomas Robert. 1798. *Essay on the Principle of Population as It Affects the Future Improvement of Society, with Remarks on the Speculations of Mr. Godwin, M. Condorcet, and Other Writers*. J. Johnson.

Mastick, G. S., R. McKay, T. Oligino, K. Donovan, and A. J. López. 1995. Identification of target genes regulated by homeotic proteins in *Drosophila melanogaster* through genetic selection of Ultrabithorax protein-binding sites in yeast. *Genetics* 139: 349–363.

Matsuo, Yoshinori, and Tsuneyuki Yamazaki. 1984. Genetic analysis of natural populations of *Drosophila melanogaster* in Japan. IV. Natural selection on the inducibility, but not on the structural genes, of amylase loci. *Genetics* 108: 879–896.

Maynard Smith, John. 1993. *The Theory of Evolution*. Cambridge University Press.

McLean, I., and F. Hewitt, eds. 1994. *Condorcet: Foundations of Social Choice and Political Theory*. Edward Elgar.

McShea, Daniel W. 1994. Mechanisms of large-scale trends. *Evolution* 48 (6): 1747–1763.

McShea, Daniel W. 1996. Perspective: Metazoan complexity and evolution: Is there a trend? *Evolution* 50 (2): 477–492.

Mendel, Gregor. 1865. Versuche über Pflanzen-Hybriden. Vorgelegt in den Sitzungen vom 8. Februar und 8. März 1865. *Verhandlungen des naturforschenen Vereines in Brünn* 4 (1): 3–47.

Meyer, C. G., J. May, A. J. Luty, B. Lell, and P. G. Kremsner. 2002. TNFa^{-308A} associated with shorter intercals of *Plasmodium falciparum* reinfections. *Tissue Antigens* 59: 287–292.

Mody, Monica, et al. 2001. Genome-wide gene expression profiles of the developing mouse hippocampus. *Proceedings of the National Academy of Sciences USA* 98: 8862–8867.

Moss, Lenny. 2003. *What Genes Can't Do.* MIT Press.

Nitzan, S., and J. Paroush. 1982. Optimal decision rules in uncertain dichotomous choice situation. *International Economic Review* 23: 289–297.

Nolfi, S., and D. Parisi. 2002. Evolution of artificial neural networks. In *Handbook of Brain Theory and Neural Networks*, second edition, ed. M. Arbib. MIT Press.

Nomes, S., M. Clarkson, I. Mikkola, M. Pedersen, A. Bardsley, J. P. Martinez, S. Krauss, and T. Johansen. 1998. Zebrafish contains two Pax6 genes involved in eye development. *Mechanisms of Development* 77: 185–196.

O'Brien, Gerard, and Jon Opie. 1999. A connectionist theory of phenomenal experience. *Behavioral and Brain Sciences* 22: 127–148.

Ohtsuki, Sumio, Michael Levine, and Haini N. Cai. 1998. Different core promoters possess distinct regulatory activities in the *Drosophila* embryo. *Genes & Development* 12: 547–556.

Paley, William. 1802. Natural Theology: or, Evidences of the Existence and Attributes of the Deity, Collected from the Appearances of Nature. Available at http://assets.cambridge.org.

Papatsenko, Dmitri, and Michael S. Levine. 2008. Dual regulation by the Hunchback gradient in the *Drosophila* embryo. *Proceedings of the National Academy of Sciences USA* 105 (8): 2901–2906.

Pirkkala, Lila, Päivi Nykanen, and Lea Sistonen. 2001. Roles of heat shock transcription factors in regulation of the heat shock response and beyond. *FASEB Journal* 15: 1118–1131.

Plutarch. 75 [2005]. The life of Alexander the Great. In *Greek and Roman Lives*, ed. A. Clough. Dover.

Purugganan, Michael D. 2000. The molecular population genetics of regulatory genes. *Molecular Ecology* 9: 1451–1461.

Raff, Rudy A. 1996. *The Shape of Life: Genes, Development, and the Evolution of Animal Form*. University of Chicago Press.

Reznikoff, William S. 1992. The lactose operon-controlling elements: A complex paradigm. *Molecular Microbiology* 64: 2419–2422.

Robert, Jason S. 2004. *Embryology, Epigenesis, and Evolution: Taking Development Seriously*. Cambridge University Press.

Rosenfeld, N., and U. Alon. 2003. Response delays and the structure of transcription networks. *Journal of Molecular Biology* 329: 645–654.

Rumelhart, David E., Geoffrey E. Hinton, and James L. McClelland. 1986. A general framework for parallel distributed processing. In *Parallel Distributed Processing*, volume 1: *Foundations*, ed. D. Rumelhart et al. MIT Press.

Saito, Takuya, et al. 2002. Analysis of monoamine oxidase A (MAOA) promoter polymorphism in Finnish male alcoholics. *Psychiatry Research* 109: 113–119.

Sansom, Roger. 2007. Legacies of adaptive development. In *Integrating Evolution and Development: From Theory to Practice*, ed. R. Sansom and R. Brandon. MIT Press.

Sansom, Roger. 2008a. Countering Kauffman with connectionism: Two views of gene regulation and the fundamental nature of ontogeny. *British Journal for the Philosophy of Science* 59 (2): 169–200.

Sansom, Roger. 2008b. The connectionist framework for gene regulation. *Biology and Philosophy* 23:475–491.

Sansom, Roger. 2008c. Evolvability. In *The Oxford Handbook of Philosophy of Biology*, ed. M. Ruse. Oxford University Press.

Schank, Jeffrey, and William Wimsatt. 1988. Generative entrenchment and evolution. In *PSA1986*, vol. 2, ed. A. Fine and P. K. Machamer. The Philosophy of Science Association.

Schlessinger, Ehud, Peter Bentley, and R. Beau Lotto. 2005. Analysing the evolvability of neural network agents through structural mutations. In *Advances in Artificial Life*, ed. M. Capcarrere et al. Springer-Verlag.

Segal, J. A., J. L. Barnett, and D. L. Crawford. 1999. Functional analysis of natural variation in Sp1 binding sites of a TATA-less promoter. *Journal of Molecular Evolution* 49: 736–749.

Sejnowski, T., and C. Rosenberg. 1987. Parallel networks that learn to pronounce English text. *Complex Systems* 1: 145–168.

Serov, V. N., A. V. Spirov, and M. G. Samsonova. 1998. Graphical interface to the genetic network database GeNet. *Bioinfomatics* 14: 546–547.

Shapley, L., and B. Grofman. 1984. Optimizing group judgmental accuracy in the presence of interdependencies. *Public Choice* 43: 329–343.

Shen-Orr, S. S., R. Milo, S. Mangan, and U. Alon. 2002. Network motifs in the transcriptional regulation network of *Escherichia coli*. *Nature Genetics* 31: 64–68.

Shin, Hyoung D., et al. 2000. Genetic restriction of HIV-1 pathogenesis to AIDS by promoter alleles of IL10. *Proceedings of the National Academy of Sciences USA* 97: 14467–11472.

Shore, P., and A. D. Sharrocks. 2001. Regulation of transcription by extracellular signals. In *Transcription Factors*, ed. J. Locker. Academic Press.

Sinha, Neelima R., and Elizabeth A. Kellogg. 1996. Parallelism and diversity in multiple origins of C_4 photosynthesis in the grass family. *American Journal of Botany* 83: 1458–1470.

Stanley, Steven. 1973. An explanation for Cope's rule. *Evolution* 27 (1): 1–26.

Sterelny, K. 2000. Development, evolution, and adaptation. *Philosophy of Science* 67: S369–S387.

Sterelny, K., and P. E. Griffiths. 1999. *Sex and Death: An Introduction to Philosophy of Biology*. University of Chicago Press.

Stern, David L. 1998. A role of Ultrabithorax in morphological difference between *Drosophila* species. *Nature* 396: 436–466.

Stern, David L. 2000. Perspective: Evolutionary developmental biology and the problem of variation. *Evolution* 54: 1079–1091.

Streelman, J. T., and T. D. Kocher. 2002. Microsatellite variation associated with prolactin expression and growth of salt-challenged tilapia. *Physiological Genomics* 9: 1–4.

Tautz, Diethard. 2000. Evolution of transcriptional regulation. *Current Opinion in Genetics & Development* 10: 575–579.

Thanos, Demitrius, and Tom Maniatis. 1995. Virus induction of human IFN beta gene expression requires the assembly of an enhanceosome. *Cell* 83: 1091–1100.

Thiessen, G., A. Becker, A. Di Rosa, A. Kanno, J. T. Kim, T. Munster, K. U. Winter, and H. Saedler. 2000. A short history of MADS-box genes in plants. *Plant Molecular Biology* 42: 115–149.

Trefilov, A., J. Bernard, M. Krawczak, and J. Schmidtke. 2000. Natal dispersal in rhesus macaques is related to serotonin transporter gene promotervariation. *Behavior Genetics* 30: 295–301.

Venter, J. C., et al. 2001. The sequence of the human genome. *Science* 291: 1304–1351.

Waddington, Conrad H. 1966. *Principles of Development and Differentiation.* Macmillan.

Wagner, Andreas. 2001. The yeast protein interaction network evolves rapidly and contains few duplicate genes. *Molecular Biology and Evolution* 18: 1283–1292.

Wang, Wen, Frédéric G. Brunet, Eviater Nevo, and Manyuan Long. 2002. Origin of sphinx, a young chimeric RNA gene in *Drosophila melanogaster. Proceedings of the National Academy of Sciences USA* 99: 4448–4453.

Watson, J., and F. Crick. 1953. A structure for deoxyribose nucleic acid. *Nature* 171: 737–738.

Westin, J., and M. Lardelli. 1997. Three novel notch genes in zebrafish: Implications for vertebrate Notch gene evolution and function. *Development Genes and Evolution* 207: 51–63.

White, Kevin P., Scott A. Rifkin, Patrick Hurban, and David S. Hogness. 1999. Microarray analysis of *Drosophila* development during metamorphosis. *Science* 286: 2179–2184.

White, Robert J. 2001. *Gene Transcription: Mechanisms and Control.* Blackwell Science.

Wilkins, A. S. 2002. *The Evolution of Developmental Pathways.* Sinauer.

Williams, G. C. 1966. *Adaptation and Natural Selection.* Princeton University Press.

Wimsatt, W. C. 1986. Developmental constraints, generative entrenchment, and the innate-acquired distinction. In *Integrating Scientific Disciplines,* ed. W. Bechtel. Martinus Nijhoff.

Wolpert, L. 1970. Positional information and pattern formation. In *Towards a Theoretical Biology,* volume 3, ed. C. Waddington. Aldine.

Wray, Gregory A., Matthew W. Hahn, Ehab Abouheif, James P. Balhoff, Margaret Pizer, Matthew V. Rockman, and Laura Romano. 2003. The evolution of transcriptional regulation in eukaryotes. *Molecular Biology and Evolution* 20 (9): 1377–1419.

Wray, Gregory A., and Christopher J. Lowe. 2000. Developmental regulatory genes and echinoderm evolution. *Systematic Biology* 49: 28–51.

Wright, Sewall. 1982. Character change, speciation, and the higher taxa. *Evolution* 36: 427–443.

Yuh, Chiou-Hwa, Hamid Bolouri, and Eric H. Davidson. 1998. Genomic cis-regulatory logic: Experimental and computational analysis of a sea urchin gene. *Science* 279: 1896–1902.

Index

Adaptive reactivity, 23, 67–69, 75
Average node perturbation effect, 60–62

Bateson, William, 11–16
Behe, Michael, 20

Canalizing functions, 49–53, 60, 63
Cell type, 42, 43, 48, 52, 65–68
Complexity
 accidental vs. expected, 37, 38, 43–46
 adaptive developmental, 1–19, 110
 evolutionary trend toward, 92–94
 in gene regulation networks, 51–53, 91
 genetic, 13–16, 53
 gradual evolution, 7–9, 18, 29, 87, 95
 irreducible, 20, 21
 and robustness, 90–92
Condorcet's jury theorem, 84
Control
 causal, 98–104
 design, 98, 102–106
 importance of understanding, 25, 97
Controller
 of development, 98, 100, 104–111
 of a process, 104, 105, 106
 of a system, 105, 106
Conway Morris, Simon, 43

Darwin, Charles, 6–10, 14, 15, 18, 21, 29, 38, 87
Darwin, Erasmus, 2, 3, 8, 14
Dawkins, Richard, 16–20, 108
Development, 2, 17, 18, 25, 26, 34–38, 43–46, 51–69, 110, 111

Evolvability, 18–22, 26, 29, 33, 34

Fisher, Ronald, 15

Gene centricism, 18, 21, 22, 108, 109
Gene expression profile, 25, 80, 84, 90, 111
Generative entrenchment, 93, 94
Gene regulation
 adaptive, 25, 80–88
 connectionist framework, 59, 68–95, 98, 110, 111
 deterministic modeling, 39–41
 evolvability, 26, 46, 69, 79, 86–94, 108, 110
 and organism development, 21, 22, 25, 26, 33–35, 86, 87, 107, 108
Gould, Stephen Jay, 17, 18, 43–46

Hurley, Susan, 98

Indicators, 77, 81, 87, 89, 95
Inheritance
 of acquired characteristics, 5, 10, 15
 adding factors, 12
 blending view, 9–12, 15

Inheritance (cont.)
gain in knowledge, 16
Mendelian view, 10–16
unknown mechanism, 9, 13
Intelligent design, 20

Jenkin, Fleeming, 10, 15

Kauffman network, 38–42, 47–53, 58–62, 88, 89
Kauffman, Stuart, 26–68, 76, 79, 80, 85, 88–90, 94, 95
Krüppel regulation network, 70, 74, 75, 81, 86–89

Lamarck, Jean-Baptiste, 3–10, 14
Leclerc, Robert, 90–92
Lewontin, Richard, 18, 21, 29
Linear separability, 54, 78. *See also* Quantitative consistency

Malthus, Thomas Robert, 7, 8
Maynard Smith, John, 108
McShea, Daniel, 46
Mendel, Gregor, 10–13
Methylation, 79
Modern synthesis, 16
Moss, Lenny, 107
Mutation
continuity, 11, 12, 18, 21, 22, 25, 29–31, 34
random, 9, 19, 28, 73, 87
somatic, 79

NETtalk, 72, 73
Network
accuracy, 67–69, 75–78, 82–88, 91, 92, 95
balance, 62–64, 67, 82–87
Boolean, 39–41, 47–51, 54, 81
complexity, 26–28, 32–35, 46, 47, 59, 80, 85, 88–95
generic, 31, 32, 38, 42, 89
learning, 71, 73, 74, 85
order, 38, 41–68, 76, 85
overfitting, 85
representation, 72–74, 85, 86

Origin of life, 2, 6, 14, 15, 44, 46, 68

Paley, William, 1, 2
Parity thesis, 109
Positional information, 70, 81, 86

Qualitative consistency, 54–63, 68, 78–84, 88, 89, 95, 110
Quantitative consistency, 54–61, 78–84, 88, 89
Quasi-independence, 18, 21, 22, 25, 29

Raeymaekers, Luc, 47, 62–64
RNA, 79
Robustness, 90–94

Selection
gene as unit of, 16–18, 108, 109
pressures, 90–92
stabilizing, 90, 92
Sensitivity
network, 60, 61, 62, 85, 88–90
microenvironment, 88, 89
Kauffman network, 61, 89
Sterelny, Kim, 19, 21

Training set, 73, 85
Transcription factor, 24, 40, 51–54, 57–59, 62, 63, 67, 68, 70, 76, 77, 93, 94, 105, 108, 110

Waddington, Conrad, 23–25, 110
Weismann, August, 15
Wray, Gregory, 33, 52, 62, 63